Fuelwood: the energy crisis that won't go away

Erik Eckholm
Gerald Foley
Geoffrey Barnard
Lloyd Timberlake

An Earthscan Paperback

© Earthscan 1984
ISBN No 0-905347-55-2

Published by the International Institute for
Environment and Development, London and
Washington, DC

Typeset by Wetherby Typesetters, Wetherby, UK
Printed by Russell Press, Nottingham, UK

This book was produced by Barbara Cheney
and Rebecca Fowler.

Cover photo: Mark Edwards/Earthscan

Earthscan is an editorially-independent news and
information service on global development and
environment issues. Part of the International
Institute for Environment and Development, it is
financially supported by the UN, the aid agencies
of the EEC, Canada, Denmark, Finland and
Norway. This book has benefited from funding
from the Swedish International Development
Authority (SIDA), the UN Food and Agriculture
Organization, the World Bank, the US Agency for
International Development and the Netherlands
Foreign Ministry, but none of these agencies is
responsible for any opinions expressed in it.

 **Contents**

Summary 5

Chapter 1: The underlying crisis 9

Chapter 2: Fuelwood and deforestation 19

Chapter 3: Getting and using wood 29

Chapter 4: Why people grow trees 47

Chapter 5: Why people don't grow trees 54

Chapter 6: Helping people grow trees: farm forestry 60

Chapter 7: Helping people grow trees: communal forestry 71

Chapter 8: Making community forestry work 82

Chapter 9: Improved cooking stoves 88

Chapter 10: Strategies for the future 99

Summary

Chapter 1: The underlying crisis

The world 'fuelwood crisis' began to receive wide attention in the mid-1970s. Since then, the situation has become steadily worse. In many developing nations, over 90% of domestic energy is supplied by wood. By the year 2000, some 2.4 billion rural people will be using fuelwood faster than it is being replenished. The poor bear the brunt of increasing shortages: their search for fuel takes longer; they must switch to inefficient, smoky fuels; huts are colder, and fewer hot meals are cooked. The World Bank and various development agencies got into the fuelwood business quickly; over 1978-83, some $500 million in aid went to community forestry. But much was done before the problem was really understood. Data are only now becoming available to explain the peculiar relationships between the woodfuel crisis and deforestation, between woodfuel supply and demand systems. It has also become clearer why many community forestry programmes have failed.

Chapter 2: Fuelwood and deforestation

In most countries, forests are disappearing because people need land for farming, not fuelwood to burn. Almost all countries deforest rapidly as they develop; Europe and North America went through such a phase. But populations are growing faster today; many Third World countries have gross inequalities in landholding. Some national laws encourage people to gain land by clearing forests. And in many countries forest is being cleared which would be more valuable for environmental protection, timber or other forest products. Fires and livestock can also wipe out forests. Urban fuelwood demands are added to these pressures. In the Sahel, the trees around the capital cities, where demand is concentrated, disappear into cooking fires. The poor are not ignorant of the process of deforestation or blind to its effects. They cut because they must.

Chapter 3: Getting and using wood

There are enormous variations in fuelwood consumption — not only from nation to nation, but also within the same village. Where it is scarce, people are already using as little as possible. In some countries, fuelwood is becoming a valuable traded commodity; but in much of Africa it is still collected freely. Where it becomes too scarce, the poor turn to dung, stalks, husks and other agricultural residues. So the fields do not get this natural fertiliser. Third World industries — tobacco curing, tea and coffee-drying, brick-making, brewing, etc — can outbid rural villagers for fuelwood. Trees are also needed for building poles and implements. Competition for wood accelerates deforestation; but it can also give people a good reason to plant trees.

Chapter 4: Why people grow trees

People have been growing trees without outside help for centuries, either as crops or along with crops (such as the trees which shade coffee plants). In countries with no traditions of planting trees, people often manage existing trees to ensure a steady supply of fuelwood. The assumption that people need to be educated into the techniques of tree growing is often simplistic and unjustified.

Chapter 5: Why people don't grow trees

There are many reasons why people do not grow the wood they need. In some areas farmers believe trees bring crop-eating birds. Some farmers lack enough land, or laws may make the ownership of trees on that land dubious. Often tree planting competes with the more important planting of food crops, or with opportunities to earn cash in off-farm jobs.

Chapter 6: Helping people grow trees: farm forestry

'Community forestry' covers various approaches to help farmers and local communities to grow trees. There are four basic types. First, in 'commercial farm forestry', farmers plant trees as a cash crop. This has worked well in parts of India where there are large wood markets. But it has been criticised for not helping the poor, or for even making their situation worse. Second, programmes encouraging 'tree growing for family use' can be helpful where there is no large wood market. The distribution of tree seedlings forms the backbone of such programmes.

Chapter 7: Helping people grow trees: communal forestry

Third, 'communal forestry' has local people and the forestry department cooperating to grow trees on public land. It has proved very difficult to organise, as even in a small village few people have similar goals and priorities. Finally, 'land allocation schemes' give people, usually the poor, the right to use land in return for growing trees. Such schemes have been used for growing wood for power stations in the Philippines; but elsewhere some types of land allocation programmes have been criticised for exploiting the landless poor.

Chapter 8: Making community forestry work

Several things must happen for community forestry to work. Forestry departments must abandon their policemen image, gain the trust of the local people and become teachers rather than guards. Non-governmental organisations, often effective where government agencies are not, can be given a bigger role and more help. Women, who are responsible for fuelwood supply in much of the world, must have a say in such programmes. Planners need to get the technical package — tree species, soil, planting regime, etc — right for little-studied areas such as the Sahel. In many regions, it might prove more effective to manage existing forests better than to plant new plantations.

Chapter 9: Improved cooking stoves

The improved wood-burning stoves now being disseminated, often by international organisations, in Africa, Asia and Latin America are designed to use less wood than the ubiquitous open fire. But because many of them deteriorate or are used wrongly, they save much less wood in the village than in the laboratory. They are unlikely to save much wood at the national level or to slow deforestation. But they can improve health and safety and general standards of living — and make it easier for individual families who desire to save fuel.

Chapter 10: Strategies for the future

Coping with the fuelwood crisis will require not one solution but a combination of activities: village woodlots, increased tree growing on private land, better forest management and the rational use of crop residues. Careful planning of plantations and greenbelts will be needed to supply cities with wood and charcoal. Where wood's true costs, both in cash and in environmental degradation, become too high, governments may want to encourage the switch to kerosene and bottled gas. But solutions must be found which include the poor. The 'fuelwood crisis' goes beyond fuelwood to become part of the larger crisis of underdevelopment. Development and social reforms which give the poor a chance to improve their lives will help relieve fuelwood scarcities.

Chapter 1

The underlying crisis

While the richer half of humankind ponders the vagaries of oil prices and the dilemmas of nuclear power, the poorer half still relies on our most ancient fuel, wood, for cooking and home heating. As wood to meet these most basic energy needs becomes scarce, hundreds of millions of Third World residents face an energy crisis of their own. By adding to the drudgery and expense of daily survival, this other energy crisis further degrades the well-being of people already desperately poor.

Until the mid-1970s the firewood crisis was largely ignored by Third World governments and international aid agencies alike. In an age of cheap oil and high hopes for global industrialisation, those who thought about the problem at all tended to assume it would disappear of its own accord. But spiralling oil prices and persistent mass poverty have revealed the futility of that hope.

More recently, firewood scarcity has become a major topic of concern. Third World energy planners have, often for the first time, taken note of the fuels most critical to as much as 90% of their national populations. Aid agencies have begun to support community-oriented tree growing and to experiment with fuel-saving cookstoves.

Heavy reliance on wood for energy is, of course, nothing new. As recently as 1850, the United States used wood for 90% of its fuel. Nor are wood scarcity and resulting hardships for the poor a new thing. Fuelwood shortages were acute in mid-nineteenth century Europe, and in an indirect way may have had epochal historical significance.

In Prussia during the early 1840s, five-sixths of all criminal prosecutions involved thefts of wood — usually by desperate peasants who poached wood off the landholdings of the rich. Young Karl Marx watched debates in the Rhineland parliament on the subject of wood-stealing penalties. He later said that it was his exposure to the human impacts of wood scarcity that first drew his attention to the material basis of social relations.

Some things seldom change. As much as in Marx's time, today's firewood crisis reflects the disparities in income and power between rich and poor. With relatively little pain the better-off can purchase what wood they need, obtain it from their own lands, or switch to alternative fuels. And governments, despite new attention to fuelwood scarcity, devote disproportionate resources to fueling urban industries and the automobiles of the affluent.

In the United States and Europe, economic growth and technical change solved the fuelwood problem by permitting large-scale shifts to new fuels. In time, economic development could solve the Third World's household fuels crisis too, although one of the key alternatives of the past — inexpensive oil — will not be available.

All indications are, however, that unaided market forces will not relieve the fuelwood problem soon enough to avert widespread human suffering and ecological disruptions. Realistic projections of population, income, and the cost of alternative fuels provide no hope that the demand for fuelwood will shrink dramatically in the coming years.

Even as the demand for fuelwood swells, the forests and other vegetation available to supply it continue, for numerous reasons, to shrink. If recent trends continue unchecked, poor people in much of Africa, Asia, and Latin America will be able to cook their meals only at tragic costs for themselves and the environment. Already, some are reduced to scavenging for bits of bark, garbage, or any other burnable materials.

Over the last decade, newly aware governments and aid agencies have made their first efforts to confront the firewood crisis. Unfortunately, if understandably, their enthusiasm has often outstripped their knowledge about local fuelwood cutting and forest trends, and the complexities of tree growing and of family cooking habits. But through a few successes and many failures, much knowledge has been gained.

Studies of rural energy and community forestry are progressing rapidly, and new details are constantly reported. Enough lessons have already been learned to permit the design of far more sophisticated responses — with better chances of making a real difference — than in the past.

In this book we draw on the last decade's experience to provide a new assessment of the fuelwood problem and possible responses. In particular, we show how intimately fuelwood scarcity is interconnected with the agricultural and other pressing land-use dilemmas of underdevelopment — and why efforts to promote tree-growing in isolation from this broader context are apt to fail.

Governments must not only do more to help people attack the fuelwood crisis; they must do it better. The alternative is deeper drudgery and dimmer prospects for hundreds of millions of people.

Dependence on wood

Wood is by far the most widely used household fuel in the developing world. In most countries, virtually every rural family relies on wood for all or part of its cooking and heating. In most Third World cities, charcoal and wood remain the predominant cooking fuels of the poor and middle classes. And the bulk of the wood that is cut in the Third World is burned in homes: a recent survey of 76 developing countries found that about eight times as much wood is used for domestic cooking as for lumber and other industrial purposes.

Because poor people cannot afford alternatives, wood remains the main fuel even in areas where forests are rapidly disappearing. In Nepal, families depend on wood for 97% of their household energy. In Upper Volta, also in the grip of severe deforestation, the dependence is over 90%.

In Guatemala, 80% of the rural population use wood as their sole cooking fuel, and a further 15% rely on it to meet part of their needs. The same pattern is repeated in the rural areas of Central America and all the poorer countries of South America. And in Africa south of the Sahara, dependence on woodfuel is almost total. In countries such as Malawi, Tanzania, Ethiopia, Somalia, Chad, Upper Volta and Mali, wood does not merely predominate over other household fuels; it provides over 90% of the *total* national energy consumption.

Country	Wood as % of total energy consumption	Country	Wood as % of total energy consumption
Angola	74%	Morocco	19%
Benin	86%	Mozambique	74%
Brazil	33%	Nepal	98%
Burundi	89%	Nicaragua	25%
Cameroon	82%	Niger	87%
Central African Rep.	91%	Nigeria	82%
Chad	94%	Pakistan	37%
Chile	16%	Papua New Guinea	39%
El Salvador	37%	Rwanda	96%
Ethiopia	93%	Senegal	63%
Ghana	74%	Sierra Leone	76%
Guinea	74%	Somalia	90%
Honduras	45%	Sri Lanka	55%
India	36%	Sudan	81%
Ivory Coast	46%	Tanzania	94%
Kenya	70%	Thailand	63%
Liberia	53%	Tunisia	42%
Madagascar	80%	Upper Volta	94%
Malaysia	8%	Zambia	35%
Mali	97%	Zimbabwe	28%

Figure 1. Fuelwood consumption estimates in developing countries.

Sources: Hall, Barnard and Moss *Biomass Energy in the Developing Countries,* (1982) Pergamon Press; Leach *et al, A Comparative Analysis of Energy Demand structures of Countries at Different Levels of Economic Development,* (1983) IIED, London.

In slightly better off countries such as Kenya, Zimbabwe and Senegal, where the proportion of total national needs supplied by wood is somewhat lower, rural people still depend almost entirely on wood. This is true even in oil-rich Nigeria.

The share of national energy consumption accounted for by wood is shown for a range of countries in Figure 1. These estimates are crude, since fuelwood consumption is notoriously difficult to measure, but the message is clear. Wood is a vital energy source throughout the Third World. Countries ignore this fact at their economic and ecological peril.

A worsening future

In 1980, the UN Food and Agriculture Organization (FAO) examined the present position and future prospects for fuelwood supplies throughout the developing world. Though isolated reports of wood scarcity and its human and environmental impacts had been surfacing from many different countries, the FAO study was the first to try to put numbers on the magnitude of the crisis worldwide. The findings were startling.

Region by region, the FAO analysts estimated the natural growth rates of existing vegetation, how much fuel could be derived from it on a sustainable basis, and the proportion that was accessible to local populations. They then matched these findings with estimates of fuelwood consumption to indicate the current balance of supply and demand. The FAO study also projected supply and demand to the year 2000, taking account of increases in population and the cumulative effect of overcutting of the remaining forest resources.

Countries and sub-regions were classified into zones according to their wood-supply position:

* *Acute scarcity* situations were defined as those 'with a very negative balance where the fuelwood supply level is so notoriously inadequate that even overcutting of the resources does not provide the people with a sufficient supply, and fuelwood consumption is, therefore, clearly below minimum requirements'.
* *Deficit* situations are those where the people are still able to meet their minimum fuelwood needs, but only by overcutting the existing resources and jeopardising future supplies.
* Other areas were classified as being either in a *satisfactory or a prospective deficit* situation.

The study found that more than 100 million people, in some 26 countries, already faced acute fuelwood scarcity. The most serious situations were identified in the following zones:

* the arid and semi-arid zones south of the Sahara;
* the eastern and south-eastern parts of Africa;
* the mountain areas and islands of Africa;
* the Himalayan region in Asia;
* the Andean plateau; and
* the densely populated areas of Central America and the Caribbean.

The study also concluded that nearly 1.3 billion people — 39% of the total population of the developing countries — live in 'deficit' areas and thus face a looming of crisis.

The FAO forward projections show things worsening nearly everywhere. Without major changes in forest management or consumption patterns, conditions in acute scarcity zones will deteriorate further. And many areas now in deficit will soon face acute scarcity. In other regions, where supplies are now satisfactory, conditions of deficit or acute scarcity will emerge. In all, by the end of the century some 150 million rural dwellers will live in conditions of acute scarcity and some 1.8 billion will be overcutting available forest resources.

As FAO admits, the data on which its study was based are often poor, and many 'informed guesses' about supply and demand trends were required. The figures conceal many of the complexities of the crisis and the great differences in the nature of the woodfuel problem in different locations. In particular, the projections cannot be regarded as accurate predictions, nor were they intended as such. The study does not take full account of the use of agricultural residues and other non-forest sources of fuel supply, and does not incorporate the changes in supply and demand patterns that are inevitable as wood prices rise.

But there is no escaping the underlying trend the study reveals: For huge numbers of people, fuelwood will be increasingly scarce and expensive. And no alternative fuels are in sight that could relieve this deepening crisis.

The human impacts of fuelwood scarcity

Like clean water and nutritious food, cooking fuel is a basic need of life. As wood becomes scarce, poor families suffer in many ways.

In many Third World cities and towns, the purchase of woodfuel is putting enormous strain on the already overstretched budgets of the urban poor. Paying more for fuel means less money for food, education, health, decent houses and other necessities. Working class families in cities such as Niamey, Ouagadougou, Bujumbura (Burundi), Port-au-Prince and some Guatemalan towns pay 20-30% of their meagre incomes for cooking fuel. Some states in India, which long ago opened 'fair-price' shops to provide subsidised food to the urban poor, have now been forced to open subsidised wood depots to enable them to cook.

In most rural areas, fuelwood is seldom bought and sold. As the vegetation around villages declines, people — usually women and children — walk farther and farther in search of wood; what was once a quick chore can occupy many backbreaking hours. Hence fuel gathering joins water collection as a source of daily drudgery, cutting into potentially productive time, especially of women.

In parts of rural Peru, women spend 10% of their time gathering and cutting wood. In northern Ghana, one full day is required to gather three days' supply of wood. Women, often with babies strapped on their backs, may walk eight kilometres (5 miles) to their husbands' bush farms to gather headloads. A recent survey in rural Kenya found that some women spend up to 20-24 hours per week collecting fuelwood. The time devoted to wood collecting varies greatly even within the same country. While women in some

Mark Edwards/Earthscan

Children in Bangladesh bring fuelwood to market. Firewood selling offers the poor and landless a chance to earn cash, but they must often gather illegally from state forests or someone else's land.

valleys of Nepal spend only a few minutes a day collecting fuel, those in other valleys spend up to 11 days per month gathering fuelwood.

Some families living near Kathmandu designate one child to spend all his or her time scrounging for cooking fuel. The amount of time devoted to wood collection is not, however, a perfect index of scarcity, for as high quality wood becomes less accessible, families are often forced to switch to nearer, lower quality fuels.

At times of the year when even crop residues are extremely scarce, landless families in Bangladesh struggle to cook with fruit skins, sugar cane wastes and other garbage — a difficult, time-consuming and smoky task. In general, the switch from wood to scavenged materials means a degradation of the quality of life. Moreover, the increased use of crop residues and dung for cooking can reduce food production, since compost made from dung or residues improves cropland quality and adds essential nutrients.

Fuelwood scarcity can threaten health. In colder areas such as the Himalayan hills, the Andes and the mountains of Central America, burning wood to keep warm in the evening is simply too expensive to contemplate; families shiver and are more susceptible to disease. Wherever wood is scarce, boiling drinking water is out of the question for the poor.

Anecdotal accounts suggest that nutrition possibly is also being adversely affected in regions of exceptional fuelwood shortage. Families in some communities in the Sahel, Ecuador and Haiti have reportedly switched from cooking two hot meals a day to only one. Some African families have reportedly tried eating ground grains in uncooked gruels. Soybeans, introduced for nutritional purposes in an area of Upper Volta, were reportedly not accepted by local women because they need too much cooking. The many possible links between cooking fuel scarcity and nutrition deserve urgent research.

Wood scarcity does not affect all people equally. It is mainly a crisis for the poorest. Those with higher incomes can switch to other fuels or pay for wood brought in from elsewhere. Well-off farmers generally have enough trees or extra crop residues on their own land to get by without hardship.

Even where trees appear plentiful, the poor may not have the right to collect wood from them. As prices rise and wood becomes a market commodity, landless labourers and tenant farmers may be denied traditional rights to collect wood or residues on large estates. In Nepal, India and Bangladesh, it is the poorest rural residents who are forced to poach for wood on public lands and incur the risk of fines or jail.

The international response

By the mid-1970s, the negative human and ecological effects of spreading fuelwood scarcity were becoming too obvious to ignore. At the same time, the recent tripling of oil prices had led policymakers to realise that the substitution of kerosene or other fossil fuels for firewood, which in the past had seemed a simple solution, was not feasible. Indeed, as oil prices soared, some communities that had long cooked with kerosene were forced back to reliance on wood.

Concern about the emerging 'fuelwood crisis' rippled through international aid agencies and affected Third World governments. Programmes to counter it were launched with a rapidity unusual in the inertia-ridden world of development planning.

The 1978 publication by the World Bank, the world's major aid-giving institution, of its *Forestry Sector Policy Paper* helped legitimise concern for fuelwood problems among sceptical economic planners and politicians. The report acknowledged that past international aid to forestry had been inadequate and too narrowly focused, and indicated that the Bank would begin to support community-based tree-planting for fuelwood and other purposes such as watershed protection.

John Spears, senior forestry advisor at the World Bank, recently summarised how the Bank and other institutions were thinking at that time:

> "In the late 1970s, after 25 years of almost exclusive focus on the industrial value of forests and extensive technical assistance and investment programmes for pulp mills, sawmills and industrial plantation forestry, it was belatedly realised that such programmes were doing little for the vast majority of the rural and urban poor of the developing world."

The new concept of 'community forestry' has been actively developed and promoted by FAO through its programme on Forestry for Local Community Development, created in 1979 with a special contribution from the Swedish International Development Authority (SIDA). The term encompasses such activities as the promotion of village and family woodlots to produce fuel, fodder, and poles; the integration of trees into cropping systems; private commercial forestry; tree planting to stabilise hillsides or shifting sand dunes; and the growing of trees along canals, railways and roads.

In the early 1970s, international aid for fuelwood and community forestry activities was negligible. What little there was came through private voluntary agencies and volunteer programmes like the Peace Corps. But between 1977 and early 1984, the total aid given for community forestry and related activities amounted to more than $500 million, with the World Bank the single largest donor. Many bilateral aid agencies and other multilateral institutions have also become deeply involved, and voluntary agencies still play key roles at the village level. More than 100 community forestry projects in dozens of countries have been funded in the last five years. In addition, large state-run plantations to help meet urban wood demands have been established in some countries.

Though the sums spent are still small compared to the size of the problem, the speed with which aid money have been mobilised is impressive, particularly given the lack of experience and institutions in the Third World for implementation. In many countries the surge in investments has had the air of a crash programme.

Efforts to increase fuelwood supplies have often been accompanied by programmes attacking the demand side of the problem by introducing fuel-efficient cooking stoves. Though improved stove programmes have received only a tiny fraction of the funds that community forestry has, about 100 projects are now under way in many countries. In Nepal, India, Malawi and several other countries, they are part of large, integrated forestry programmes attacking the problem from all sides. Improved stoves are also being developed and promoted by voluntary and non-governmental agencies.

Results to date

The results of the programme to date have been mixed. Large-scale tree-planting projects, some successful, some not, have been pursued in China over the last quarter century. In South Korea, a massive reforestation effort has been successfully carried out through Village Forestry Associations. India has recently initiated 'social forestry' campaigns with some success in establishing village woodlots and much more success in encouraging private farmers to grow trees, though questions have been raised about the impact of the latter on the rural poor.

Some community forestry programmes in other parts of the world, notably in the Sahel, have been almost total failures. In the worst cases, such as a project in Niger, seedlings were actively destroyed by local people because they had been planted against their will on traditional grazing grounds. Few stove programmes have yet been successful — whether judged in terms of getting large numbers of new stoves into kitchens or of saving significant amounts of wood.

Learning from experience

With their perception of the very existence of the fuelwood problem so new, governments and aid agencies rushed to confront it in great, if understandable, ignorance. In most of the wood-short countries, planners knew little about the state of forests and the actual role of fuelwood collection in their decline; about how much wood families use and where they obtain it; about local traditions of forest management and why they have proved inadequate; and about why people have not acted on their own to plant more trees.

Both foreign funders and national officials often assumed that rural people saw fuelwood scarcity as a top-priority problem and, given financial and technical assistance, would gladly cooperate in planting trees. A widespread belief that fuelwood collectors were major agents of deforestation, with its many ill effects, lent a sense of urgency to the new activities.

Considering the dearth of data and experience on which they were based, it is not surprising that many community forestry efforts have failed to achieve their targets. But, in good part because of these pioneering programmes, much valuable knowledge has been gained. We have learned, for example, that the nature of the fuelwood problem varies tremendously from one place to the next — making local participation in planning not just an idealistic goal but a practical necessity.

We have learned that the fuelwood problem is just one strand in the whole cloth of natural resource management, only one of many interlinked resource challenges facing rural people. We have learned that fuelwood collection is just one of several causes of deforestation, and is seldom the major one. We have learned that planting and raising trees is no easy task, and requires as much attention to social organisation as to silviculture. In sum, we have learned that the fuelwood crisis is indeed real and a threat to the welfare of the world's poor — but that both the problem and its potential solutions are quite complex.

In less than a decade, a large and growing literature about the fuelwood crisis has appeared, and valuable new lessons are being learned each year. The danger is that the disappointing results of many recent forestry programmes will cause support for future ones to flag. All evidence indicates that worldwide efforts to counter the fuelwood crisis must be redoubled even as they are redirected. If the emerging lessons can successfully be applied, then the failed programmes will have served their purpose after all.

Chapter 2

Fuelwood and deforestation

The rampant loss of forests now occurring in many Third World countries is without doubt a serious threat to their economic development. The direct economic costs become apparent as forest products of all kinds become scarce and expensive. Over time, the ecological impacts may exert an even greater price. As hillsides are unwisely cleared, the incidence of deadly landslides and flashfloods rises, and heightened sediment loads choke reservoirs and canals downstream. The decline in tree cover in semi-arid zones encourages the loss of topsoil and the spread of desert-like conditions. Ill-planned clearance of rainforests causes an irreversible wastage of potentially valuable biological resources without any sustained economic gain in return.

Deforestation and the firewood crisis are obviously closely linked. But the two problems are not identical. To understand and effectively attack fuelwood scarcity, it is crucial to understand what is happening to local forests and why, how forest trends affect the availability of fuelwood, and how trends in fuelwood use affect forests.

An image has often been drawn of rising human populations stripping adjacent forests bare of trees, thereby creating fuelwood scarcity. Except around certain cities, that image is not usually accurate. More often, recent research has revealed, fuelwood scarcity is as much a consequence as a cause of deforestation. First, the widespread clearing of lands for agriculture severely reduces the available forest area. At that point, the gathering of fuel from the remaining woodland may well begin to exceed the sustainable harvest.

Much of the overcutting is not in forests at all. Rather, rural people gather their fuel from trees and shrubs scattered around farms and throughout the countryside. People generally prefer to collect dead wood. But as the ratio of people to trees rises, fuelwood collecting may, along with heavy grazing, help deplete the vegetation over large areas and cause the replacement of useful plant species by useless ones. This is not deforestation in the sense of forest clearing, but it can be as much a threat to ecological stability and local prosperity.

Whatever its links to deforestation in a particular place, the fuelwood problem is important in its own right. Families need fuel for cooking and heating; an extreme scarcity of wood can impoverish their lives and their surrounding environments. Fuelwood collection is just one element in the cycle of poverty and bad planning that produces runaway deforestation in a country. A broader land-use strategy to save forests and a programme to meet fuelwood needs must go hand-in-hand.

From forest to farm

In most countries, forests are disappearing not because people want the trees to burn, but because they want the land under the trees for agriculture. For government agencies sponsoring colonisation, agribusinesses establishing plantations or pastures, and landless peasants seeking a family plot, getting rid of trees and wood, not growing more, is the immediate problem.

Not all deforestation is bad. Most of the world's best cropland was once forested. Where potentially arable land exists under forest and people need it to grow food, then converting forests to agriculture can make economic and social sense. But much forest conversion is ill-planned and both economically and environmentally unsound. Often the environmental protection, timber and other forest products a forest offers outweighs the benefits that can be gained by clearance for agriculture.

Europe and North America learned these lessons the hard way in times past. Though it is hard to believe today, much of Greece was once clothed in forest. But denudation around Athens was already so severe by the 4th century BC that Plato lamented: 'Our land, compared with what it was, is like the skeleton of a body wasted by disease. The soft plump parts have vanished, and all that remains is the bare carcass.'

Much of the dense forest that once covered Western Europe was cleared for farming or was logged during the Middle Ages. But the rampant clearing of mountain slopes in the Alps and Pyrenees produced a severe ecological backlash. Alpine torrents and landslides were chronic threats by the end of the 18th century, prompting reforestation and forest protection. Today, about a quarter of Western and Central Europe is forested and the forest area, much of it intensively managed, is stable — or at least was stable until air pollution and acid rain recently began to damage some forests.

Settlers rapidly deforested North America. Much of this clearing was sensible, but some of it was self-defeating. By the early 20th century, with one-third of its original forests gone, the US government established a national forest system to preserve forests for ecological, economic and recreational purposes. More intense forms of agriculture and the growth of non-agricultural employment allowed forests to return to marginal, often heavily eroded croplands.

Almost all countries go through a period of rapid deforestation at early stages of their development. Today, however, developing countries face unique pressures that make the transition to a sustainable balance of human settlements, forest and farmland extraordinarily difficult.

Population growth rates are higher than those faced by Northern countries as they developed, and the human increase often outpaces the creation of non-agricultural jobs that could reduce the demand for new farmland. Gross inequalities in landholdings, together with farm mechanisation and rapid population growth, force the rural poor to farm steep hillsides, drought-prone areas, and rainforest lands that ought to remain under tree cover for ecological reasons. Unlike the earlier cases of Europe and North America, most Third World countries have no new worlds to colonise, and no vast, fertile empty spaces that can serve as geographic safety valves.

Even with modern techniques, many tropical forest soils are unsuited to permanent or intensive agriculture. Once cleared, the fertility of forest soils can rapidly decline as nutrients previously tied up in the vegetative layer are leached away, and as wind and water erosion set in. Once productive forest land can be transformed into unusable wasteland in short order.

Clearing forests to make way for agriculture is an established government policy in many countries. The results have been mixed. The establishment of productive oil palm and rubber plantations in western Malaysia has shown how conversion from rainforest to agriculture can be successful, given a careful choice of sites and appropriate technical and financial support. The tree crops protect the soil from the damaging effects of concentrated sun and rain.

Mark Edwards/Earthscan

The world's closed forests are being felled not for firewood, but to clear the land for crops, pastures and settlements. Here settlers are clearing land for farms after a new road cut through the Peruvian rainforest.

In Indonesia's transmigration scheme, which aims to move millions from crowded Java and Bali to the so-called outer islands, some of the colonies have prospered while others have been disastrous for both people and land. The latest $500 million World Bank loan for the scheme is largely for rehabilitation of the areas where forest colonisation has gone wrong.

In Latin America, legal and tax incentives have encouraged forest clearance. In the 1960s and 1970s, the government of Brazil offered large tax incentives for the establishment of cattle ranches in the Amazonian rainforest. This was also backed by large foreign investments. For four or five years, cattle ranching on ex-rainforest land is highly profitable. But then the grasses give out and the land often has to be abandoned.

In some Central American countries, people can gain legal land rights simply by moving into forest areas, clearing the trees and planting crops. Their claim to the land becomes stronger the longer they farm it.

This leads to 'two-stage' deforestation in Costa Rica. Poor farmers clear the forest and cultivate the land for a few years. When fertility begins to decline, the farmers sell their squatters rights, usually to rich cattle owners, and the land is turned over to grazing. Forestry officers in the area claim they can detect the intentions of squatters by observing their homesteads: those who intend to stay often plant fruit and fodder trees around their homes for their own uses.

Much forest clearing today is unplanned and simply beyond the control of governments. It is a result of land hunger among poor, displaced people with no alternative.

When new logging roads provide access to remote forest areas, a rush of settlers often follows. Through this process, according to FAO, nearly 600,000 hectares (1.5 million acres) of forest were cleared annually between 1976 and 1980 in the Ivory Coast and Nigeria alone.

In India, southern Nepal and elsewhere, landless people move into forest reserves, clear a plot and hope that political pressures and sympathy will prevent their eviction. In the Philippines, four to five million people — equivalent to 10% of the national population — are illegally occupying public forest land.

Nepal's southern plains, known as the Terai, were covered with dense forest until malaria was eradicated there in the 1950s. Migrants from Nepal's overcrowded hills (and some from India) rushed to settle the area. It made sense to convert the Terai, with its rich land, to mainly agricultural use. But the chaotic colonisation process has resulted in the total denudation of large areas and severe shortages of fuelwood and timber. Ironically, Nepal has now been forced to ask the World Bank to provide funds for a reforestation programme in the Terai.

Where the soils are suitable for farming, careful land clearance and adoption of appropriate farming techniques can allow permanent, productive settlement on forest soils. But often the squatters who move into forest areas lack the skills necessary to manage the land sustainably. Nor do they have the funds needed to practice modern agriculture. And those who have no legal rights to the land have no incentive to invest in permanent land protection measures. Often, then, the land is farmed to exhaustion and left degraded, useless either for agriculture or for productive forestry. Squatter families are then forced to move deeper into the forest.

In the end, the protection of essential forests depends on policies and trends outside the forestry sector per se. The pressures to convert forest to farmland must be reduced by maximising employment and food output on existing farmlands; by developing rural, small-scale industries to provide non-agricultural jobs; and by maximising labour use in the industrial sector. In general, development policies that spread assets and work more widely among the poor, and that contribute to a decrease in population growth, will also serve to reduce land hunger and slow deforestation. Governments too often allow or even promote forest clearing as a less painful alternative to social reforms elsewhere. Such an approach may have short-term political appeal, but is at the expense of future prosperity.

Livestock and fires

In moderate numbers, grazing animals cause little serious damage to forests. But too many animals damage trees and stunt their growth by continually removing foliage; often trees are stripped so heavily that they are killed. Animals also keep forests from regrowing by eating or trampling the tender shoots and seedlings.

Cyprus provides a dramatic example of the long-term effects of overgrazing. Goats are excluded from graveyards, so these areas have more trees and grasses than the adjacent land. As a consequence, graveyards sometimes tower above their surroundings, from which several metres of soil have been removed by wind and water erosion. Similar sights can be seen in Ethiopia.

In many areas, livestock grazing and fodder gathering by people cause more damage to woodlands than does fuelwood cutting. John Wyatt-Smith, a British forester, demonstrated this in a study of the Phewa Tal catchment area in Nepal, where livestock provide draught power and dung for fertiliser. About a quarter of the animals' fodder comes from the forests around the villages, much of it in the form of leafy branches that farmers cut from live trees. Wyatt-Smith found that 'the area of forest land required to sustain the fodder needs of the farming system is about three to five times greater than that required to provide fuel and timber needs'.

Similar conclusions were reached by American researcher Jeff Fox after detailed analysis of the Bhogteni village area in the hills of Nepal. He found the livestock population to be three to nine times higher than the local land could support, and wrote:

> "...grazing was found to pose the greatest threat to environmental stability, followed by tree fodder collection. Contrary to prevailing belief, firewood collection was not a serious threat to the degradation of Bhogteni's public lands... The undue emphasis placed on firewood as the cause of deforestation has obscured the role played by livestock."

Overgrazing by animals helps to cause deforestation and desertification in many drier regions. As populations of humans and their animals grow, cropland is expanded and bigger herds must make do with smaller rangelands. The resulting environmental degradation, particularly around waterholes and along migration routes, can be severe, as has been noted in many of the semi-arid grassland areas in East Africa, the Middle East, the Sahel, and parts of Latin America.

Both nomadic herdsmen and settled farmers often deliberately start fires to clear the ground so that when the rains come a fresh crop of grass, fertilised by the ashes, grows quickly. Fires are often used in hunting or to clear new

Mark Edwards/Earthscan

Cutting twigs from a living tree to get fuelwood in northern India. Trees can stand a lot of this, but often roadside trees are overcut and die.

lands, and sometimes spread accidentally too. In Honduras, pine stands are burnt to help control cattle ticks.

After repeated burning, only fire-tolerant tree species — usually smaller and much slower growing — will remain. This may finally turn forest land into much less productive savannah or grassland. Fires, both deliberate and accidental, have been cited as an important cause of forest decline in sub-Saharan Africa.

Fuelwood for rural needs

In rural areas, fuelwood cutting is rarely the cause of large-scale deforestation. But as the forests recede, pressure on the remaining trees grows. Over time, fuelwood scarcities can develop, with fuelwood cutting becoming a major factor in the devegetation of the countryside.

People living near forest reserves may gather their wood in them, legally or illegally. But most of the firewood villagers burn comes from trees outside designated forests. These trees, dotted around the farming landscape, provide building poles and animal fodder as well as fuel. They are also sources of fruit, herbs, medicines, fibres, leaves and fungi. Though wealthier rural people may be able to afford modern alternatives, the poor often rely on trees for a whole array of vital daily needs.

How fuelwood cutting affects non-forest trees has rarely been studied; neither their presence nor their decline is usually included in forest statistics. But on the basis of available information we can identify some common stages in the evolution of rural fuelwood scarcity.

Where local population densities remain low, the demand for fuelwood can usually be met without damaging the local standing stock of trees. By travelling short distances from their homes, people are able to gather sufficient wood without damaging the resource base. People who own land may be able to make do from their own trees. Where shifting cultivation is practised, clearing trees from fallow land will often yield more than enough wood for domestic uses.

In some areas, local custom permits people to collect dead wood from their neighbours' lands. Wood may be gathered from common land, roadsides and wastelands. For people who live within walking distance of government forest reserves — probably a minority of the rural population in most countries — fuelwood may also be available from these areas depending on rights of access.

Where fuelwood is plentiful, usually only dead wood is collected. It is lighter to carry than green wood, it is easier to cut and it burns better. Though branches may be lopped off, whole trees are rarely felled merely to provide fuel.

As rural populations grow, the increase in fuelwood demand is usually met at first by extending the collection area. People search longer and travel farther to obtain good quality fuelwood. But provided there are enough trees in this wider area, supply and demand can remain roughly in balance; the cost is in terms of productive time, usually women's.

Eventually, people will not be able to find new sources of wood within walking distance. It is then that fuelwood cutting can start to cause deforestation, as fuelwood demands begin to eat into the remaining standing stock of trees. At this point, people begin turning to lower quality wood, roots, crop residues, dung and other combustible materials.

The cutting of live branches becomes more common. Too much of this can kill trees. In parts of India, the once majestic shade trees along roadsides are being damaged and killed by overcutting, both for fuelwood and for animal fodder. When wood becomes even scarcer, people may cut a ring of bark around the tree's trunk to kill it. When the tree has died and dried out they come back and chop it down.

But the cutting of trees is seldom uniform or indiscriminate. Some species, such as the 'pipal' tree or sacred fig *(Ficus religiosa)* in Nepal, are never cut for fuel as they are sacred and are protected by taboos. Similar religious or traditional restrictions apply to cutting neem and banyan trees in parts of India, while the baobab tree is left standing in many parts of Africa long after neighbouring trees have been cut.

Trees on private land usually survive longer than those on public land since their owners try to protect them. In Kenya, for example, the traditional practice of allowing neighbours to collect fuelwood from private land has become rarer as wood becomes scarcer. Traditional communal wood-collection areas may become severely depleted, or they may be taken away by private individuals, a phenomenon noted in Mali, Upper Volta and Niger.

Such restrictions force people, particularly the poor, to rely more on remaining communal resources, speeding their depletion. Appearances can be deceptive. Even in areas that seem relatively well stocked with trees, the poor may face a wood crisis because they have no right of access to available wood.

In such a deteriorating situation, fuelwood cutting helps deplete the remaining trees, but it is not the only factor. As rural populations grow, so do livestock herds and so do the demands for cropland and for poles and other building materials.

Where shifting cultivation is practised, increasing population forces farmers to shorten the traditional cycle of cultivation and fallow. Shorter fallow periods mean less time for shrubs and trees to regenerate. Wood becomes scarcer and the vital role that trees play in restoring soil fertility is undermined. In huge areas of semi-arid Africa, such a decline in traditional fallow periods is both reducing the tree cover and destroying the fertility of food-producing lands.

The impact of urban and industrial fuelwood demands

Urban and industrial consumers, unlike rural villagers, are a cause of much outright deforestation. Unable to gather wood themselves, and often unable to afford fossil fuels, many city dwellers have no choice but to buy wood or charcoal that has been carried or trucked in from rural areas. Many Third World industries such as tea and tobacco curing, brick-making and coffee drying purchase enormous amounts of fuelwood, with no accountability for the environmental impacts of their business.

If people can sell wood or charcoal, they have a much stronger incentive to cut live trees. This happens near many Third World cities, where both big businessmen and powerless peasants see a chance to profit by selling wood in a commercial market. Such woodcutting is seldom regulated. In Africa's semi-arid zones, an expanding ring of severe depletion surrounds cities like Dakar, Ouagadougou and Niamey.

Near Kharagpur in the Indian state of West Bengal, an area formerly covered with a rich sal *(Shorea robusta)* forest has been reduced to scrub. Gradually, despite intense governmental efforts to protect the forest, people have cut it down to sell wood in the towns. The devastated area now stretches to the horizon in all directions.

In some countries large well-organised gangs of wood cutters, operating illegally or in collusion with forest guards, plunder protected forests. But often very poor people are forced by their grim circumstances to poach wood for sale in cities. A study team from the Indian Institute of Management in Bangalore described the syndrome of desperation:

> "...for a villager who has no food, the attack on forests is for collection of firewood for sale in urban and semi-urban centres, rather than his own consumption, because selling firewood is often the only means of subsistence for many poor families. This firewood, with the help of bus and truck drivers, goes to the urban markets like Bangalore... Theft of wood as a means of survival is becoming the only option left for more and more villagers. Recently 200 villagers were caught stealing firewood in the Sakrabaile forest of Shimoga district and one person was killed in a police encounter."

Lured by the urban market, farmers may sell off wood from their own lands. In times of hardship, farmers may be forced to cut their live trees even though they realise they are endangering their own future wood supplies. Investigators from Bharathidasan University found that in some villages in Tamil Nadu:

> "...distress sale of trees, because of drought conditions, is reported. This indicates that the villagers resort to short-term exploitation of fuel

resources in drought periods when their incomes fall drastically, unmindful of the long-term consequences of their act."

Around the city of Kano in Northern Nigeria, over the last 25 years commercial wood demands have led to severe deforestation and the collapse of a sustainable agricultural system. Once farmers would lop branches from the trees on their land during the dry season and transport them into the town on donkeys to sell in the market. While in town, they picked up dung and sweepings from the streets to use as fertiliser on their fields. But rising fuelwood demands in Kano encouraged farmers to overcut trees, selling off their biological capital; now farmland within a 40-kilometre (25-mile) radius of the city has been largely stripped of trees. Charcoal-making, almost always a commercial activity, is often a significant cause of forest destruction. Because it is lighter and easier to transport than wood, charcoal can be trucked hundreds of kilometres to markets, thereby extending the reach of urban-induced deforestation. The inefficiency with which traditional artisans convert wood into charcoal only adds to the damage.

Deforestation is extensive along the floodplain of the Senegal river, where once extensive stands of *Acacia nilotica* have been cut for production of charcoal that is shipped to markets by boat. Elsewhere in Africa, too, better roads have resulted in accelerated forest destruction as urban charcoal markets become accessible to more remote rural areas.

In Haiti, with a big urban fuel demand and little remaining forest, charcoal production may be the single largest cause of deforestation. Live trees are cut, as opposed to the dead branches and twigs that provide the bulk of rural fuelwood supplies. Charcoal production is carried out only by the very poor. The response of local people has been summarised by USAID consultant Frederick Conway:

> "Local residents in all of the research sites recognised deforestation as a great problem. Deforestation is seen as contributing to floods and drought. Even young adults can remember when the hillsides, now denuded, were covered with trees. Furthermore, charcoal production is perceived as the cause of this deforestation. More to the point, poverty is seen as the cause of deforestation because only poverty leads a person to make charcoal. Rather than resentment against charcoal-makers as destroying a natural resource, there is great sympathy for such people."

In sum, urban fuelwood demands are often an important source of deforestation, not only of adjacent lands but also, where charcoal is used, of far away woodlands. The commercial pull of urban markets also adds to the pressures on the trees of the countryside.

Chapter 3

Getting and using wood

In the last few years, many fuelwood surveys and detailed village energy studies have been conducted. Gaps still exist, but a clearer picture of the fuelwood economy is emerging, which should enable planners to establish more effective programmes in the future.

Rural fuelwood consumption

Fuelwood consumption varies enormously, both between and within countries. A survey of 518 small farm families in different parts of Nicaragua, where most cook with fuelwood, showed that annual consumption ranged from 1,100 to 2,865 kg (2,400-6,300 lb) per person, almost a threefold variation.

Not surprisingly, the availability of fuelwood is a key determinant of how much people use. In areas of the Third World where it is plentiful, people may use as much as 2,000 kg (4,400 lb) per head per year or more. Where wood is scarce, people may use only 500 kg (1,100 lb) per head. Where scarcity has forced people to switch to fuels such as straw and animal dung, fuelwood consumption may drop to near zero.

In villages in Lesotho and in South Africa's Transkei, where wood is scarce, the annual per capita fuelwood consumption was 288 and 271 kg (635 and 597 lb), supplemented by 260 kg and 80 kg (570 and 180 lb) of dung, respectively. But in a village in KwaZulu in South Africa where wood is abundant, the consumption was 1,124 kg (2,478 lb) per head. And here the climate is much milder than in Lesotho and the Transkei.

The village of Kwemzitu in the mountains of northeastern Tanzania is close to a forest reserve from which the people are allowed to take their fuel. The walk takes about an hour but the supply is not otherwise restricted. The annual wood consumption was found to be in the range 1,636-2,605 kg (3,607-5,743 lb) per head.

In much of the Sahel, wood is scarcer and consumption lower. Surveys of six villages in Mali and Niger found that people used only 440-660 kg (970-1,455 lb) per head per year. The survey report notes that where wood is scarce people switch to non-preferred tree species and try to economise. Nevertheless, in this area, the figure of 440 kg (970 lb) appears to be a floor below which people find it difficult to reduce their consumption.

Similar consumption figures come from different wood-short areas in India. Seventeen villages in Tamil Nadu showed a range of 344-676 kg (758-1,490 lb) per head, with an average of 481 kg (1,060 lb). Fuelwood consumption in five villages in the state of Orissa ranged from 509-826 kg (1,115-1,821 lb) per person, averaging 680 kg (1,500 lb).

Family size influences per capita fuel requirements. Large families, cooking bigger meals, can be much more efficient in their use of fuel. In one village in Nepal, families with between one and four members used an average of 890 kg (1,962 lb) of firewood per person each year. Families with nine members or more used less that half this amount per person — about 340 kg (750 lb).

Consumption often varies during the year. In this same Nepalese village, consumption is highest in January, the coldest month, and in October, when wood is needed for festivals. In November and December, after the harvest, fuelwood use drops considerably because corn cobs and crop stalks take its place.

Funerals, weddings, and feasts can consume huge amounts of wood. In India, up to 400 kg (880 lb) may be used to cremate a single body. In Gujarat, the forest department is giving villages improved crematoria that reduce consumption by 25-50% and thus help people save both money and trees.

Collecting the wood

In most countries it is the women, perhaps helped by children, who are responsible for fuel collection. Thus, not surprisingly, in some areas men seem unconcerned about fuel scarcity problems and indifferent to suggestions that they should plant trees.

But as wood becomes scarce, collection methods and responsibilities can change. Up to the early 1970s, in one village in the central province of Papua New Guinea, fuelwood collection was done by women near their homes or on their way home from farm gardens. As shortages emerged, men became involved. Groups of 5-10 men and women now hire trucks and travel 30 km (19 miles) to a mangrove forest. The men chop down the trees and cut them into logs, while the women split them into the smaller pieces required for firewood.

In the Mbere region of Kenya, firewood collection, like most tasks, once involved a strict division of labour based on sex. Today, partly because of scarcities, these divisions are breaking down, according to a study by US sociologists David Brokensha and Bernard Riley. Previously, "a circumcised boy could not carry firewood at all: it was 'mugiro', a prohibition, and an offender would have difficulty in marrying. But today many have to carry firewood." Here, as in many other regions, men are becoming increasingly involved in firewood collection, though it is still predominantly a women's task.

Emerging commercial wood markets

In the rural areas of most developing countries, people collect the wood they need without paying money for it. But in many places this traditional free fuelwood gathering is giving way to commercial transactions.

This can result both from the increased scarcity of wood and from wealthier people's desire to avoid the trouble of collecting firewood. The poor often turn to collecting firewood and selling it to the rich as a way of earning cash. Also, the impacts of a commercial fuel trade may be spreading outward from the cities.

Commercialisation has reached different stages in different countries. In Malawi, only 7% of rural families buy any fuelwood, and of these only 1% rely on bought wood entirely. In Tanzania, only salaried public servants such as teachers or rural extension officers purchase their fuelwood. These families make up less than 2.5% of the rural population.

Mark Edwards/Earthscan

Women in Ayorou, Niger, bargaining fiercely over wood prices. Most urban people must buy their fuelwood, and as prices rise farmers may be more tempted to plant trees for the city markets.

Mark Edwards/Earthscan

These women of the Ethiopian highlands are carrying sugar cane stalks home to burn. Where crop residues are scarce, burning them for fuel robs soils of nutrition and may accelerate erosion rates.

But 65% of households in the village of Ibb, in the Southern Uplands region of the Yemen Arab Republic, get at least a quarter of their fuelwood from the market. In Nicaragua, as much as 60% of the country's woodfuel consumption may be commercialised.

The development of a commercial woodfuel market in an area can radically change the way local people regard the question of woodfuel supplies. When people can sell wood, there will be a cash incentive for them to grow trees if this is the most profitable use of their land. But there is also a greater incentive to cut them down, either from their own land or from communal areas and nearby forest reserves. Thus the spread of cash markets for fuelwood is at once a potent cause of forest depletion and a potential force for remedial action.

Malian anthropologist Jacqueline Ki-Zerbo has described how West African men's interest in fuelwood rises as it becomes a commercial commodity:

> "The behaviour of a man who uses a motor bike, a cart or even an automobile (for transporting wood) is significantly different, since for him bringing home wood is no longer considered a sign of inferiority and submission to his wife. On the contrary, it is the recognition of

Mark Edwards/Earthscan

Bangladesh has few forests left, and dung accounts for about a quarter of traditional fuel use. This farmer makes 'fuelwood' by wrapping dung around palm frond stalks; the result can be fed slowly into simple stoves.

a new situation, that is that wood has become a rare and therefore valuable commodity deserving the attention of the male sex.''

Whether this newfound interest can be tapped on behalf of forest renewal remains to be seen.

Also, when people have to buy wood, they have a much stronger incentive to conserve it. Investing in a more efficient stove makes more sense if people can pay back the cost of the stove through fuel savings. And there is a stronger reason to switch to other fuels. For the better off this may mean a switch to 'modern' fuels such as bottled gas, kerosene or electricity. For the poor it often means turning to lower grade fuels such as straw and dung.

The commercialisation of a formerly free good can pose special dangers for the poor and landless. In parts of India, landless families who once had the right to collect wood freely on landowners' farms have lost this privilege as landowners see a chance to sell their excess wood for cash.

Agricultural residues: the fuel of the poor

Where fuelwood is scarce or expensive, poorer families often cook with

This Nepalese woman uses rice straw to fuel a two-pot mud stove.

agricultural residues: dried dung, straw, stalks or other vegetable matter. Researchers are finding that the burning of such residues is far more widespread than previously assumed.

In some places, residues are used exclusively all year round. Large areas of Bangladesh no longer have a fuelwood crisis; few trees are left, and most people cook with straw and dung. Elsewhere, residues and wood are alternated or used together. In southern Nepal, people roll a mixture of rice husks and dung around a stick, combining three types of fuel at once.

Often, the degree to which residues or wood are burned varies with the seasons. Villagers in central Upper Volta cook with millet husks during the half-year they are plentiful, and with wood during the other six months.

Cooking with residues is an ancient practice in China, parts of which were deforested thousands of years ago. The nature of Chinese cuisine, with many dishes that are fried quickly over high heat, may well be explained by the need to cook over the brief, intense flames of straw fires.

Though data are scarce, the burning of residues and other biomass materials seems to be increasing as fuelwood supplies become short in more regions. Studying this phenomenon in Indonesia, Paul Weatherly has called it 'energy involution':

> "...where in one year a family or an industry mostly used stemwood grade firewood, the next year they use coconut husks and bamboo roots. Still later, rice hulls, corn stalks, tobacco stalks, rice stalks and finally dried leaves are commonly used as cooking fuels. In a completely involuted system, no burnable bit of biomass is immune from inclusion in a cooking fire."

The switch to residues and other available organic materials may mean that people spend less time collecting fuel than before. In some villages in Orissa, the main forest has receded so far that villagers can no longer travel to it, so they have turned to collecting brushwood from around the village. Similar patterns have been noted in Malawi.

Such a shift expands the amount of fuel available in an area. But there are costs. Less convenient and flexible for cooking, and often less aesthetically pleasing to handle and burn, these lower grade fuels can reflect a lowering of the quality of life. They may take less time to collect, but they may make more work for the cook, who must take greater care in feeding and tending the fire.

But some people prefer crop residues to wood. Elizabeth Ernst found that in the villages in central Upper Volta she studied, "the women prefer to use, as much as possible, millet stalks instead of wood". Whereas gathering wood required a four and a half hour journey and much laborious chopping and carrying, millet stalks could be collected as needed near their homes. To slow the burning, women wet the stalks bit by bit as they push them into the fire.

Charcoal-making in Haiti. A pile of wood, a covering of straw and dirt; it wastes energy, but this is the way charcoal is made throughout the Third World.

Urban fuelwood markets draw wood supplies from the rural areas and hasten the depletion of local wood resources (Dhaka, Bangladesh).

Since dung is often used as a fertiliser and soil conditioner, its increasing use as a cooking fuel has generated widespread concern. Dung has long been a major household fuel in India, Pakistan and Bangladesh, and is used to a lesser degree in some parts of Africa, the Middle East and the Andean zone of South America, where llama dung is burned.

Statistics on the extent of dung burning are notoriously poor. The FAO made a rough estimate in 1979 that about 400 million tonnes of dung were burned annually in Asia, Africa and the Near East. How much of that total would otherwise have been applied to fields is not known, but a tonne of dung is believed to have the potential to raise grain yields by about 50 kilograms per hectare (45 lb/acre).

Where applying dung to fields is not practical, such as in grazing areas traversed by nomads, burning it for fuel probably does no harm. For 19th century settlers in the Great Plains of the United States, dried 'buffalo chips' were a convenient, costless source of household fuel.

Hindus consider the cow a sacred animal. Together with the bullocks, buffaloes and other animals that roam the streets of Indian towns, they provide an important source of fuel for the poor. Throughout Calcutta and other cities, dung cakes can be seen drying in the sun, neatly laid out on pavements or stuck on walls, trees and even lampposts. Even if it was not burnt, it is unlikely that this dung would be returned to the fields.

Where wood must be purchased but dung is available free or at low cost, using dung in the fireplace rather than the field can make economic sense for the individual farmer. One researcher in India calculated that a tonne of cow dung applied on the farm would result in increased grain production worth about 80 rupees ($8). The same tonne burned for cooking would eliminate the need for firewood that costs 268 rupees ($27) in the local market. Hence, growing more firewood in a region will not necessarily mean a commensurate fall in dung burning; this will depend on relative prices.

Yet in some areas, dung and residues remain linchpins of the agricultural system, and their increased use for cooking could have devastating consequences. In the densely populated Himalayan hills of Nepal, for example, it is rarely feasible to import commercial fertilisers. Composted dung, crop residues and forest leaves are crucial to the maintenance of soil fertility. Any significant use of dung and crop residues for fuel could hasten the unravelling of a food production system that is already in deep trouble.

In Bangladesh, the use of crop residues for cooking has increased with the human population, but the supply has not. American researcher John Briscoe has described the result:

> "Given the inflexible requirements for cooking fuel, animals are fed less and are fed inferior fodders (such as water hyacinth), and the amount of organic matter returned to the land is reduced. The consequences are that the animals are unable to plough as well, and

the fertility of the soil is reduced, meaning that crop yields continue to fall."

The very word 'residue' is misleading, for in traditional farming systems virtually all the organic produce of a farm is useful in some way — as fodder, compost, fuel, thatch or fencing.

In some places the poor may have trouble getting even crop residues, and may have to either scavenge for garbage to burn or spend part of their meagre incomes on fuel. As residues become scarce and hence valuable, they may go on sale like fuelwood. Landless labourers and sharecroppers may lose traditional collection rights.

In Bangladesh, reports Briscoe:

> "Competition for the organic materials produced by the land has become intense; the number of village trials arising from disputes over the ownership of trees and crop residues is large and growing. The marginal social and economic groups are denied access to organic materials on which they previously depended for fuel and are forced to purchase fuel from the market."

Changes in agriculture can change fuel-supply patterns. Traditionally grown deep-water rice in Bangladesh has long stalks that are widely used for fuel. The new high-yielding rice varieties have short stalks and thus accentuate the fuel crisis as they increase food output.

Whether cooking with residues harms or benefits a given society can only be determined by analysis of the particular situation and alternatives. Where agricultural output is high and biomass by-products are plentiful, cooking with residues may well be a sensible answer to the household fuels problem. But the current and potential uses of agricultural residues in cooking, and the impacts on agriculture, require far more scientific study, as these are among the least understood elements of the Third World's household fuels crisis.

Urban consumption

Most Third World cities rely on wood or charcoal to supply at least part of their domestic fuel needs. In many African cities, charcoal is the most popular fuel, with wood used only by the very poor. In Latin America, wood is the dominant fuel. In Asia the picture varies: Bangkok almost exclusively uses charcoal, while Madras and most other Indian cities cook mainly with wood.

Data on urban woodfuel consumption are even scarcer than those from rural areas. Official records of charcoal entering Dakar in 1978 indicated an average annual consumption of 100 kg (220 lb) per head. In Kenya, urban charcoal consumption ranges from 100-170 kg per head per year (220-375 lb).

One estimate puts annual consumption of charcoal by urban dwellers in Tanzania at about 170 kg (375 lb) per head, and another at 315 kg (694 lb) per head. Studies of urban households in Malawi show that the annual urban consumption is about 470 kg (1,040 lb) of fuelwood and 170 kg (375 lb) of charcoal per person. The estimated annual consumption of fuelwood in Ouagadougou, Upper Volta, has been given as 438 kg (966 lb) per head.

How towns get wood

Urban people get wood by buying it. Middlemen often play an important role by buying wood and charcoal from rural suppliers and bringing it into cities by truck. In large cities, the trade is often organised around a series of wholesale depots, from which smaller retailers obtain their supplies. In Madras, these depots also provide supplies for villages surrounding the city. Trucks loaded with fuelwood often pass each other, one entering the city while the other leaves.

In some countries, powerful urban merchants or syndicates control the

Mark Edwards/Earthscan

Camels hauling wood into Addis Ababa. Across Sahelian Africa from Ethiopia to the Atlantic, strings of camels bring tonnes of fuelwood daily into the major cities, often after days on the road.

A Cairo charcoal-seller. Charcoal — a form of wood which is easier and cheaper to transport — is the main cooking fuel of most African cities.

wood and charcoal trade to cities, and they can be a major hidden obstacle to the introduction of new sources of woodfuel. In Ouagadougou, Upper Volta, the woodfuel trade is well organised and controlled by urban businessmen and high government officials. The wood is delivered by truck and by donkey carts that make journeys of several days. In the Sudan, forest rangers have accosted armed crews filling trucks with illegally cut wood for conversion to charcoal; shoot-outs have occurred.

Illegal tree-cutting can be big business. Contractors working on government or private lands have been known to falsify records and forge permits in order to cut officially protected forest stocks. In many countries there are allegations of widespread collusion between contractors and local forestry officials to evade restrictions on cutting in forest reserves.

In Kenya, truck owners control the charcoal market. They buy the charcoal from rural producers as the drivers return from delivering other goods to outlying districts, and then sell it through their own outlets in the cities. Since they do not have to make special trips to pick up the charcoal, the economics of its transportation are completely altered, and much wider areas of collection are opened up. Thus charcoal may come into Nairobi from as far away as the Sudanese border, 600 kilometres (375 miles) to the north.

Much of the supply to Dakar, the capital of Senegal, comes from licensed groups of charcoal-makers working in designated areas of forest that were killed by the droughts of the 1970s. Supplies are also brought in by truck from the Casamance area some 300 km (185 miles) away in the extreme south of the country.

Charcoal comes into the Tanzanian capital, Dar es Salaam, from the open woodlands along the roads leading into the city. It is made in small earth kilns by itinerant charcoal-makers, the majority of whom are unlicensed. They take their product to the roadside, where it is picked up by dealers and brought into the city by truck.

In smaller towns, the woodfuel trade tends to be more informal. Rural suppliers may themselves bring fuel to the towns, using donkeys or bullock carts, carrying it on buses, or bringing it on their heads. Some sell to dealers, while others trade directly in the marketplace.

Around the town of Bara in central Sudan, farmers who are clearing their land make charcoal from the cut wood in simple earth kilns. They bring the charcoal 16-18 km (10-11 miles) into town by camel.

Former nomads, displaced by drought, also sell wood in Bara and other nearby towns. As one of them admitted to Norwegian researcher Turi Hammer: "We take trees belonging to other people. We cut them when they are too young. We never pay any tax. We must live from something. What else can we do?"

Each day in Kathmandu a steady stream of people, men and women of all ages, trudge in carrying heavy baskets of wood on their backs. They have set out before dawn, walked several hours to nearby hills and cut wood, often illegally, to sell in the city market.

Whether they are on the low rungs of a syndicate or doing small-scale selling on their own, huge numbers of the poor in many countries get much of their income from urban wood sales. Thus, unless it is carefully planned, a scheme to supply cheap wood to a city could undermine the welfare of many of those it is intended to help.

Forest departments are major suppliers of urban woodfuel in some countries. In Kerala, southern India, wood from state forests was until recently sold through government depots at controlled prices. This supply has dried up due to reductions in the rate of felling and replanting in official forest reserves. Now rubber trees on private plantations are supplying much of the local needs. After about 15 years, rubber trees are subjected to a few years of rapid tapping, called 'slaughter tapping', then cut. Wood from rubber trees now finds a ready market as fuel at twice the previous official price.

Forest clearance for agricultural settlement projects can also provide substantial supplies of urban woodfuel, even resulting in artificially low wood prices. Cities that rely on such temporary sources may come to feel that cheap wood is a permanent phenomenon, while the low prices discourage tree planting to meet future needs.

Large areas of forest in Sri Lanka are being cleared as part of the Mahaweli irrigation scheme. A newly established private company has started a charcoal-making project to make use of the waste wood. The plans are to encourage small-scale charcoal production using portable metal kilns. Annual production is projected to rise to 100,000 tonnes in 1985.

Nicaraguan consumers are temporarily benefiting from several different major new wood sources. A coffee-renovation project involving the felling of shade trees and replanting of coffee plants is releasing large amounts of fuelwood and charcoal onto the urban market. Elsewhere, the expansion of agriculture following the land-reform measures of the Sandinista government is producing a similar result. When these one-time sources dry up, urban fuel prices are certain to climb and, unless tree plantation plans are soon laid, wood shortages will emerge.

Industrial fuelwood demands

Rural industries, including tobacco curing, tea and coffee drying, brick-making, beer brewing, sugar-making and others, often consume large amounts of woodfuel. In the towns, restaurants, tea shops, bakeries and laundries also add to the demands for fuelwood and charcoal.

The Third World tobacco industry, which usually relies on wood for curing, is a particularly voracious consumer of forests. In Tanzania, drying the tobacco crop from a hectare of tobacco requires the felling of a hectare of the adjacent savannah woodland every year.

The fuel requirements of brick-making depends on the quality of clay, its moisture content and the degree to which the clay itself burns or gives off heat in the baking process. In Tanzania, about 35 cubic metres (1,200 cu ft) of wood are needed to make the 25,000 bricks that go into a family house. Tea drying requires one cubic metre (35 cu ft) of wood to cure 150 kg (330 lb) of green leaves. Brewing the local beer (made from maize and other cereals) requires one cubic metre of wood per 400 litres (90 UK gal) of beer, and a typical village beer shop makes 800 litres (175 UK gal) per day.

Certain industries could not survive without free or low-cost wood. If they had to pay for their fuel, or bear the costs of planting trees to meet their needs, they would no longer be competitive in the international markets where they sell their products. This is true of both the tea and tobacco industries in some countries.

Other local industries must buy wood locally. Wood for brick-making in southern Thailand, for example, is purchased from owners of rubber plantations who are felling trees prior to replanting. In Costa Rica, wood for coffee drying is bought from nearby farmers.

Local industries can account for a high proportion of total national wood consumption. In Kenya, industries and small commercial enterprises consume as much as 23% of all woodfuel used. In Central America, industries are estimated to account for 18% of the total woodfuel demand.

G. Foley/Earthscan

A coffee-drying plant in Costa Rica. Rural industries such as coffee and tea-drying, tobacco curing, brick-making and sugar-making can consume large amounts of fuelwood.

Mark Edwards/Earthscan

A large proportion of the trees cut for the market go to provide poles. They are used to build traditional rural houses or urban shanties (Haiti, above) and to provide scaffolding for the construction of modern urban buildings (Madras).

G. Foley/Earthscan

Building poles and other demands for wood

Although most wood used in the developing countries is burned for fuel, wood also meets a wide range of other basic needs for which there are no readily available substitutes. These alternate sources of demand exert a heavy influence on fuelwood supplies and markets.

Poles, especially for house construction, are widely used and are often in scarce supply. In rural areas, wooden poles are used in many traditional dwellings. They are often employed as rafters or for walls that are then coated in mud or covered with thatch or woven mats. Larger poles are needed for centre or corner posts, or where additional strength is required. Short poles are used for tool handles, fences, parts of carts and other implements. Wooden ploughs requiring large pieces of hard, durable wood are still widely used.

The urban poor use poles to build huts and shanty dwellings, and builders need them for scaffolding and props. Cut square, they are used as roof beams and structural members in permanent buildings.

Sawmills are also important wood consumers, as are rayon and match industries. In some countries, including India and the Philippines, these industries are prepared to buy wood from smallholders. Elsewhere, there may be a 'latent' demand for wood from industry that would rapidly expand if wood were available at economic prices. Newsprint is rationed in India, and many pulp mills are running well below capacity because of the lack of pulpwood. These potential demands may encourage farmers near pulp mills — at least those who can afford the necessary inputs — to take up commercial tree growing.

Interacting demands

The various markets and demands for wood interact in ways that must be taken into account when planning fuelwood programmes. In some places, industrial demands dominate the whole wood market and push up fuelwood prices as a result. In the Kolar district of Karnataka, farmers are being offered a cash advance to grow eucalyptus. According to a study by the Indian Institute of Management:

> "The price of eucalyptus offered to the farmers is above the price of firewood. This ensures that eucalyptus wood is supplied to the few paper mills and one rayon mill in the area.... The consumers of the products of the pulpwood-based industries are urban elites and have almost infinite purchasing power compared to the rural poor."

Building a new road into an area, or improving an existing one so that buses and trucks can use it, may be all that is required to allow nearby cities or industries to outbid people of that area for local wood. So a road built for 'rural development' may have a harmful effect on the rural poor unless they are given an opportunity to share in the growing and felling of wood or other products for the newly accessible market.

The spread of commercial wood markets increases competition for local wood resources and, when not accompanied by public regulations and support for sustainable forestry, accelerates deforestation. The higher buying power of people in the towns tends to push up local prices, thus squeezing poor people out of the market and forcing them to turn to other fuels.

But it can also have positive effects on the local system by creating earning opportunities — both in cutting and selling firewood and, if the price is high enough, in growing trees. Wood produced in this way will not necessarily be sold locally; farmers will sell it to the highest bidder. Thus the products of local plantations may be sucked into the cities without helping to meet local demand. Simply getting trees into the ground will neither automatically slow deforestation nor solve local fuelwood problems.

Chapter 4
Why people grow trees

Farmers in most countries have been planting trees for thousands of years. They plant trees for fruit, fodder or nuts, for shade or to protect other crops, or simply for ornament or social prestige. Even where tree planting does not take place, there are often well developed traditions of forest management.

Trees play integral roles in traditional rural life. As Jeffery Burley of the Commonwealth Forestry Institute has observed:

> "All detailed studies of uses and perceptions of trees by rural peoples show that there is an extensive ethnobotanical knowledge, with a keen appreciation of species' properties, and that trees are used for a wide variety of purposes..."

Building on traditional knowledge and uses of trees is often one of the most effective ways of stimulating new tree growing.

Tree crops

The most obvious reason to grow trees is to make money. Trees yield a variety of valuable products including fruit, nuts, oils, spices, fibres, medicines, and gums as well as wood for lumber, poles, pulp or fuel. The rubber, coconut, coffee, date and oil palm industries are vital to the economies of many countries. These trees are not confined to large commercial plantations; they provide cash incomes for many smallholders and marginal farmers as well.

Some trees yield valuable by-products. Coconut palms provide copra for sale, and the fibre from the husk around the nut is used by small matting, rope and sack industries. Waste products such as the shells, husks and leaf stems can supply a substantial proportion of the local household fuel needs.

Rubber trees provide substantial quantities of fuel when they are felled at the end of their useful life. Coffee plant prunings are used as firewood; fruit trees, too, can provide fuel through regular pruning and trimming without affecting their productivity.

The gum arabic tree *(Acacia senegal)*, a small leguminous (nitrogen-fixing) tree that grows extensively in semi-arid Africa, fills many roles in the community. Most trees grow from naturally established seedlings, but they are carefully tended by local farmers and play an important part in restoring soil fertility during the fallow period of the traditional shifting cultivation system. The gum, tapped during the dry season, is sold for use in food and textile production, medicines and the preparation of paints and printing ink. Gum arabic has been traded for more than 4,000 years and is a major export of the Sudan.

In addition to restoring soils and providing export income, the gum arabic tree supplies protein-rich foliage and pods for camels, sheep and goats. The seeds are dried and eaten as vegetables. The hardwood is used for agricultural implements as well as for firewood and charcoal. A strong fibre can also be obtained from the tree's long surface roots, of use for rope, fishnets and well linings.

Traditional agroforestry systems

'Agroforestry' means growing trees in close association with crops, either in rotation with them or interspersed among them. Though the term has only been coined in recent years, the association of trees and crops has a long history. The Sahelian use of gum arabic trees as part of the shifting cultivation

G. Foley/Earthscan

Eucalyptus is often planted along the boundaries of fields in Gujarat, India. Most farmers in this area sell the wood to nylon and paper mills.

cycle is only one of scores of examples.

In Costa Rica, the traditional method of coffee growing involves two separate levels of trees. The lower-level trees are heavily pruned twice a year but are allowed to grow back to shade the coffee plants during the crucial period of flowering and fruit formation. Prunings are left in the fields as green manure. In addition, a few taller timber trees are grown in each field, and harvested on a 30-year cycle.

Multi-storey coffee cultivation is also practised in the East African Highlands, as on the slopes of Kilimanjaro and Mount Meru in Northern Tanzania. Here, coffee is grown with bananas and beans under shade provided by tall timber species.

One of the most highly developed forms of traditional agroforestry is found in the 'home gardens' of Southeast Asia, especially Indonesia. In small plots around their homes, families tend with great sophistication a multi-layered mixture of species that provides food, fodder, timber and many other goods. Home gardens play a particularly important role in the densely populated and intensively cultivated islands of Java and Bali; they cover 20% of Java's total arable land. Without these ecologically complex plots, Java would be in the throes of acute fuelwood scarcity and malnutrition would be far more widespread.

Less sophisticated forms of tree growing around homes and on farmlands are carried out in many other countries. Almost every family in the Tanzanian mountain village of Nyandira, for example, has a private tree plantation, and most individuals claim to have personally planted between 100 and 1,000 trees. Seedlings are normally obtained by transplanting them from natural sites. The species include black wattle, eucalyptus, and various fruit trees. Researcher Margaret Skutch writes that "this village was classified by the local forestry extension officer, who had never visited it, as a 'failure' in afforestation terms, since it had not requested or received seedlings from the Forest Division's nursery."

Even in Nepal, whose deforestation has become legendary, many families plant trees around their farm plots. In the hills, households own an average of 28 trees each. Most of these trees are grown by natural seeding, or by transplanting seedlings from other parts of their land. In Malawi, large numbers of farmers plant trees, mainly for poles, relying mainly on seedlings they sprout themselves or collect from under nearby trees.

A survey in the Fatick region in Senegal revealed that families plant an average of 50 trees each of species such as eucalyptus, neem and a variety of fruits and nuts. About half the families raise seedlings in their own family nurseries, and virtually all protect or transplant seedlings found growing naturally.

Some 48% of farmers interviewed in the Valle Occidental of Costa Rica said they planted trees as windbreaks. Trees are used extensively for live fencing throughout Central America. The leguminous species *Gliricidia sepium* is particularly popular for this purpose; even quite large stakes will sprout and grow when planted along a fence line. These live fences also provide fuelwood and a protein-rich cattle fodder.

The thorn tree *Acacia albida* is highly valued in the Sahel, being either deliberately planted or left standing in fields when they are cleared for agriculture. This nitrogen-fixing tree's extraordinarily useful role was described in a 1981 FAO report:

> "Its roots go mostly straight downwards rather than sideways, drawing up nutrients and using water that would otherwise be lost to local production. It provides shade for cattle in the dry season but sheds its leaves in the rainy season when agricultural crops are growing, thus providing them with humus and not competing with them for nutrients; and it produces poles, fuelwood, and fodder for local needs."

In semi-arid Gujarat, mesquite *(Prosopis juliflora)* is widely used for live fencing around farm lands. It can also be cut and used to protect young tree seedlings of other species from animal browsing. In Tamil Nadu almost all households cultivate some trees.

The spread of the neem tree in the Sahel shows how rapidly people can accept an introduced species when it meets their felt needs. Neem grows

Mark Edwards/Earthscan

A nomad camp in Niger. Trees are essential for providing shade for people and livestock in hot countries such as Niger, a use often overlooked by development planners.

quickly, coppices well, produces excellent quality firewood, timber and poles, and provides good shade throughout the year. It was introduced into West Africa early this century by the British and French colonial forest services for roadside planting and woodlots. Over the past four decades it has spread throughout the whole Sahel region and is now one of the more common species, particularly around settlements. The demand for seedlings is high, and private nurseries sell neem seedlings around some cities.

The grafted mango is now showing a similar process of spontaneous dissemination. Demand for these fruit trees is so high that at some times of the year they are supplied to dealers in Upper Volta from Bamako, Mali, a distance of 550 km (340 miles).

Trees are often planted to produce pods or leaves for fodder, as is the case with the thorn tree *Acacia nilotica* in parts of India. *Leucaena leucocephalia* (giant ipil-ipil) is traditionally grown by small farmers in the Philippines for backyard cattle fattening. The tree is planted along fences and boundary lines and family members cut the foliage and bring it to the cattle sheds.

Trees are sometimes grown for spurious reasons. Many farmers and even professional foresters believe that tree planting will increase rainfall. Though huge forests, such as in the Amazon Basin, can affect regional rainfall patterns, there is no scientific evidence that small-scale tree cutting or planting can alter the weather. Many people — foresters again among them — believe that trees are good because they 'produce' oxygen. But in fact, over the whole span of a tree's life, including its decomposition or burning, trees consume as much oxygen as they produce.

Producing wood for the market

In some countries trees have long been grown to produce wood for the market. Usually this has been done on a fairly modest scale, and for products more valuable than fuelwood.

In Costa Rica and Ecuador, many farmers plant a few trees around their homes and in their fields as a form of long-term insurance. These are cut and sold for building timber when money is needed for a wedding or some other large cash outlay. In Turkey, trees are traditionally planted on the birth of a female child as a sort of down payment toward her wedding costs.

In the Indian state of Kerala, landholdings are often measured in hundredths of an acre and consist of little more than a strip of garden around the dwelling. Nonetheless, many trees are crammed into these small plots. A few timber trees are often planted along with coconuts, cacao and other cash crops. Even slow maturing species such as teak and mahogany are sometimes grown as a long-term investment.

More popular in Kerala, however, is the species *Ailanthos malabarica*, which is in great demand from the match-making industry and can be harvested after 8-10 years. A mature tree currently fetches the equivalent of $40. For many years the match industry in southern India has largely drawn its raw material from smallholder tree growers and some companies distribute free seedlings to stimulate a steady supply.

Growing trees to produce fuelwood for the market is less common, but some historical and current examples exist. The most dramatic is from Ethiopia where, in the late 1890s, the Emperor Menelik faced an extreme scarcity of wood around the new capital of Addis Ababa. The surrounding landscape was so denuded of trees that some feared the capital might have to be moved. In response, Menelik exempted from taxation land planted with trees, and arranged for the distribution of seedlings at nominal prices.

The seedlings were mainly eucalyptus varieties, introduced around that time by Europeans. The programme started slowly, and by 1906 there were only a few scattered clumps of eucalyptus to be seen. In the ensuing years, however, many landowners planted several hectares in the expectation of high profits, and forests sprang up around houses and roads. By 1920, the streets of Addis Ababa began to look like clearings in a forest, and some suggested that the city's name be changed to Eucalyptopolis. By 1964, eucalyptus plantations covered 13,500 hectares (33,400 acres) around Addis Ababa.

Private plantations of *casuarina* (Australian pine) are the major source of both firewood and building poles for the Indian city of Madras. These trees have been planted as a commercial crop in coastal districts for more than 40 years. They are grown on a 4-7 year cycle, mostly on salty soils. Farmers usually sell the trees to brokers, who organise cutting and transport to Madras over distances of up to 150 km (100 miles). Similar plantations exist around Bangalore. Planting for the fuelwood market has also been reported in Indonesia, especially in the hills near Yogyakarta and in the vicinity of the densely populated rice-growing areas.

Sustained management of trees for fuelwood

Where people do not plant trees for fuelwood, they sometimes manage them to ensure a steady supply. Where trees are regenerating fast enough, people normally only collect dead and fallen wood. Elsewhere, there are strong traditions of coppicing and pollarding certain species.

Coppicing involves cutting the tree down to its stump and allowing it to regrow; normally a number of shoots replace the original single stem. Pollarding is the technique of cutting off the crown of the tree, leaving it to send out new branches from the top of the stem. Both coppicing and pollarding can be effective methods of obtaining a sustained yield of small-diameter wood from stands of trees over a long period of time.

Villagers in many regions have also mastered the fine art of lopping branches, for use as fuel or fodder, off tall trees without killing them. Danish forester Gunnar Poulsen has described in detail the technique used by farmers on the slopes of Mount Kenya. The tree used is silk oak from Australia. On a typical farm, seedlings are planted in single lines or spread across fields. About five years after planting, the trees reach a height of some 10 metres (33 ft), and start causing problems for the farmer because they compete with crops for water and shade out sunlight. At this point, writes Poulsen:

> "The farmer has a simple solution to this which enables him, in fact, to kill three birds with one stone. He will prune the trees severely, removing all branches until nothing is left of each tree but the naked stem, looking rather desolate like a flag-pole without a flag. In this way he will reduce competition for water and light almost to zero and he will at the same time obtain an always welcome supply of small wood."

In spite of this severe treatment, the trees sprout fresh foliage surprisingly quickly 'from bottom to top of the denuded stem, making them look like giant bottle brushes'. The crown also regrows, and after a few years the tree is pruned again. This may be done 15 or 20 times in 50 years. The trunk will continue to widen; it will also increase in height, unless this is deliberately prevented by pruning it at the top. Whenever the farmer decides it is large enough, or that he wants the money, the trunk is felled and sold for timber.

The lifetime contribution of a tree cropped in this way may be much greater than if it is simply allowed to grow and then is cut. Yet the potentials of coppicing, pollarding, and lopping techniques have received little research attention. This is one area where foresters may have more to learn from farmers than the other way around.

Chapter 5

Why people don't grow trees

Rural people are perfectly capable of recognising fuelwood scarcities; most know how to grow trees and recognise the benefits of doing so. Then why don't they grow more trees? Any of a variety of constraints may prevent them from doing so.

Some constraints can be overcome relatively easily. But overcoming others requires major changes in institutions, attitudes, laws and land tenure systems. Recognising and devising means of countering the barriers to tree planting are crucial first steps in any community forestry programme. Where these steps are skipped, as they often are, programmes may fail completely. In some cases the obstacles may be insuperable, and efforts to promote community forestry may be pointless.

Antipathy to trees

For cultural or entirely practical reasons, people in some places do not want more trees around. In parts of Africa, the normal method of controlling tsetse flies has been to cut down the trees that harbour them. This approach has been promoted through government-sponsored education and extension work over a long period, and people see trees as a threat to their well-being.

Elsewhere, farmers object to trees growing near their crops because they provide a haven for seed-eating birds. Seeing large flocks assembling in the trees around his fields, a farmer may be reluctant to save standing trees, let alone plant more.

Trees are widely believed to lower the water table. Where they replace vegetation that is less water-demanding, this may actually be true. Under some circumstances, trees can also compete with adjacent crops for nutrients and sunlight, thus reducing yields. Where these influences are real, farmers may justifiably feel that the benefits of trees are outweighed by their disadvantages.

Certain trees are associated with malign spirits or taboos. Tamarind trees are thought inauspicious in certain parts of India, and those close to a village are often suspected of housing evil spirits. So, though the trees provide edible fruit, people are not willing to plant them. People in some areas of India also regard trees as potential hiding places for robbers.

Lack of incentives

Even farmers who do not object to trees may have little incentive to plant them, especially in areas where they can easily get the wood they need. In many parts of the Third World, fuelwood and other tree products can still be obtained without great effort, legally or illegally, from communal land or nearby forest reserves. That their cutting will eventually lead to forest depletion may appear to the farmers as either totally beyond their control or irrelevant within the time scale in which they plan.

Perceptions can vary among individuals in the same area. A survey of local attitudes in Costa Rica revealed that farmers' views depended on whether or not they themselves had trees on their land. Those with no trees thought there was a severe regional shortage of wood, whereas those who owned trees did not feel that deforestation had reached a critical stage.

Tree planting may simply be low on peoples' list of priorities, a situation revealed in many surveys. Based on her work in the Sahel, US sociologist Marilyn Hoskins wrote:

> "In discussing basic needs in the community, people mentioned that there were things more urgently needed than forestry products. They stated that unless they could have water, health care, education for their children, jobs for young adults, and enough food and income to keep their families together, it did not matter if they planted trees for the future."

Land tenure

Land tenure problems can be a more fundamental barrier to tree growing. Often, it is only people who have secure title to land who are willing to make the long-term investment required for tree growing. As with any economic activity, people are not apt to work hard at something unless they are fairly certain they will reap the benefits.

For one reason or another, many Third World people do not have clear-cut ownership of the lands on which they live and work. This greatly complicates the tasks of tree planting and management. Marginal farmers who move onto steep slopes or logged areas are often there on an illegal or semi-legal basis, with a constant threat of expulsion. (On the other hand, in some Latin American urban shantytowns, people with no real security of tenure plant trees almost in a symbolic act of defiance against an uncertain future.)

Tenant farmers cannot be expected to plant trees unless they are sure that they will ultimately profit from their investment. Tenant farmers and landless labourers make up a large share of the rural populations of Asia and Latin

America; many are effectively ruled out from individual tree growing.

Even on privately owned land, traditional grazing rights may conflict with tree growing. In many countries, animals are allowed by custom to roam freely over everyone's land during the dry season. This makes the protection of saplings extremely difficult; by fencing off young trees the planters would infringe upon the rights of others. Where nomadic tribesmen pass through at certain times of the year, such problems can be particularly severe.

In many traditional African cultures, land is owned on a semi-communal basis. Instead of having permanent possession of a plot, individuals are granted temporary rights by the village leaders to farm particular areas. Often the land is reallocated every few years, undermining any individual interest in tree growing.

Questions of tree ownership

Even where farmers have full title to their land, the ownership of trees may be in question. In parts of the Sahel, farmers are unwilling to grow certain valuable tree species because they are on forest departments' lists of protected species. To harvest such trees, farmers have first to prove that they themselves

Mark Edwards/Earthscan

Finding communal land suitable for tree growing can be a major problem for community forestry projects. Around this western Sudan village, goats roam freely and graze on all vegetation.

have planted them, and then go through the laborious business of obtaining the necessary cutting permit.

Haitian peasants have similar fears about their rights to exploit the trees on their land. In a number of tree planting projects the peasants were told by officials that the trees belonged to the government and that they would be punished if the trees were cut down. The aim was to ensure that proper care was taken of the trees, but the reverse was more often the result. The trees were seen by the peasants as a threat that could even lead to expropriation of the land; rather than nurturing them, the peasants wanted the trees to die.

In Pakistan, small farmers have been reluctant to allow the forest department to plant trees for them on their private lands. They feared that this would lead to the government taking possession or control over their land, or that they would be deprived of fodder and grazing rights. Most did say, however, they were prepared to offer small plots for planting provided they received convincing assurances that the Forest Department would not take their lands, and that they would still be able to cut grass for their cattle. Here as elsewhere, a long legacy of hostile relations between forestry officials and farmers must be overcome.

Many national laws concerning forests and trees are directed towards the preservation of trees, with little allowance for their positive uses. In the Dominican Republic, Honduras and some other countries, ownership of all the trees is vested in the government. There are penalties for cutting any trees without permission, even those standing on a peasant's own land. The intention is to preserve forest cover and protect trees against indiscriminate cutting. But with such laws on the books, it is hard to persuade farmers to plant trees.

The Philippines also have laws, mainly directed at the wood industry, to control the cutting of trees. The process of getting a cutting permit is slow and cumbersome. As a result, some small farmers who have invested in tree growing are finding it difficult to harvest and sell their own trees.

Shortages of land

Forest departments often have a hard time finding decent land on which to plant trees. In densely populated areas, most of the cultivable land tends to be occupied by farmers. Seemingly empty areas are in fallow as part of someone's cropping cycle or are used for grazing. Officials trying to find land for communal tree growing programmes are usually told that none is available, or they are pointed to barren areas where it is difficult to get anything to grow.

As one Gujarat forester admitted, "Most of the lands we're planting are called grazing lands, but in reality they are nothing more than exercise grounds that have scarcely produced a blade of edible grass in years". Establishing a high-performance plantation on such poor land is often difficult.

But when individual farmers want to plant more trees on their own holdings, lack of land rarely seems to be a major constraint. Farmers can almost always find space for a few trees if they want to, even when their landholdings are small.

Under a current Haitian tree growing programme, seedlings are issued only to those prepared to plant a minimum of 500 trees. As there is a ready market for charcoal and other wood products, peasants can expect to earn a good return from their trees. According to the former programme director, US anthropologist Gerald Murray:

> "Our initial hesitations about posing a 500 tree minimum have disappeared in most regions, where the peasants themselves have found space for these trees and have in several instances expressed regret that they had not requested more."

In Malawi, few farmers feel that land shortages are a significant constraint on their tree planting. As a national survey concluded:

> "Of people who had not planted trees, only 18% mentioned shortage of land as a reason. The problem was cited most often in the Southern Region, where 25% of non-planters complained of land shortages. However... in the same region, 56% said that they could get enough land to plant individual woodlots if they chose to do so. Apparently, land scarcities are not now a serious barrier to tree planting in the Southern Region, and the problem is even less acute in the rest of the country."

In most places farmers will be able to find room for at least some trees if they can see any personal benefits in doing so. On the other hand, where the constraints on planting are great and the benefits unclear, land shortage may well be seen, or cited, as an obstacle. And certainly, in more densely populated countries, large areas of unused, high quality land are seldom available for tree plantations.

Seasonal competition for labour

Competition for labour can hamper tree planting, especially in arid areas. In moister zones, tree growing is comparatively easy and planting times are less exact. But in dry areas such as the Sahel, tree planting is hard work, particularly where the planter must break through a subsurface layer of hardpan so that tree roots can go deep enough.

Mark Edwards/Earthscan

Peasants in the Ethiopian highlands plant pepper seedlings. When faced with a choice of whether to plant crops or to plant trees, most small farmers must choose to give their time to their food crops.

Often the planting season for both crops and trees lasts only a few weeks each year; understandably, farmers choose to spend that critical period planting food crops rather than trees. Even when digging machinery is available, the shortage of time may prevent farmers from taking advantage of it. Thus outside labour may have to be hired if seedlings are to be successfully put in the ground.

Seasonal shortages of labour have constrained reforestation in Tanzania, according to Director of Forestry E.M. Mnzava:

> "Tree planting always coincides with agricultural activities. And naturally the latter get priority. Moreover, the women who are the fuelwood managers are also so much involved in other activities that they have little time to devote to tree planting. For instance, in Singisi village in the Arusha region women work for about 15.5 hours per day."

Tree planting and protection may conflict with opportunities to earn off-farm income. Particularly in Africa, a large part of a community's workforce leaves for employment elsewhere during the slack season. If people stay behind to weed plantations or to protect trees against grazing animals or fires, they miss out on cash earnings.

Chapter 6

Helping people grow trees: farm forestry

The terms 'community forestry' and 'social forestry' are commonly applied to the recent efforts to involve individuals and local communities in tree growing — to take forestry outside the forests. The use of these labels does not, however, necessarily mean that activities carried out in their name always bring social benefits to the community as a whole. In forestry, as in food production, simply growing more may not relieve the deprivation of those on the bottom. Who does the producing, who pays the costs, and who captures the benefits are crucial considerations.

Because community forestry on a significant scale is so new, few projects have yet produced trees ready for harvesting. But much has already been learned about the various approaches tried. There are five general types of community forestry, with the distinctions based on the land tenure under which they occur and the incentives that are used: commercial farm forestry, tree growing for family use, communal forestry, land-allocation schemes for the landless, and natural forest management. (The first two, which involve planting on private lands, are discussed in this chapter, the others in Chapter Seven.) Many of the best programmes combine elements of all these approaches.

Commercial farm forestry

'Farm forestry' is the name usually given to programmes to encourage tree growing for the market by individual farmers on their own land. FAO has described it as "turning peasants into entrepreneurs and producers". Because it relies on the cash incentive, it is obviously restricted to areas where there is a commercial market for tree products, or where a market can be created as part of the programme.

The most successful farm forestry programmes have been in India. Pioneering work carried out in Gujarat state in the early 1970s provided a model that has been adopted in other states. Seedlings are distributed to farmers, either free or at a low price, using a decentralised system of nurseries. This is coupled with publicity campaigns using radio, public figures, schools and voluntary organisations.

Farm forestry is proving extremely lucrative for some Indian farmers. Due to the high market price for wood, especially building poles and pulpwood, profits from growing trees on irrigated and fertilised land are often considerably higher than those from food or other cash crops. In the case of intensive eucalyptus cultivation, as is practised by some large-scale farmers in Gujarat, the financial return can be four times that of the highest paying agricultural crops.

Such high profits have made tree growing increasingly popular among farmers. In Gujarat, the demand for seedlings has risen sharply. In 1971, the annual distribution of seedlings was 6.1 million; in 1982 this had risen to 95 million. Preliminary estimates for the 1983 planting season put the total number distributed at around 200 million. In the Uttar Pradesh project, seedling production had to be greatly increased to cope with the demand; the total distributed to farmers by the end of the 1982-83 season amounted to 156 million, nearly 30 times the original target.

A similar farm forestry scheme has been used to encourage tree growing in Haiti. Once farmers were convinced that they would own and derive the benefits from any trees they grew, a substantial number came forward to receive free seedlings. In 1982, the first year of the project, 1.75 million seedlings were distributed.

In the Philippines, a more comprehensive incentives package has been used to encourage farmers to supply wood to a pulp plant, owned by the Paper Industries Corporation of the Philippines (PICOP), at Bislig Bay on the island of Mindanao. Here the fast growing, deciduous species *Albizia falcataria* is being used. PICOP supplies seedlings to farmers at cost price and gives technical advice.

Through a special scheme with the Development Bank of the Philippines, loans are given to cover the initial investment and to provide farmers with an annual income while the trees are growing. There is a guaranteed market with an annually reviewed minimum price for the pulpwood produced. Considerable areas have been planted, but, following typhoon-caused tree damage in 1982, many farmers fear they will be unable to repay their loans. Farmers have also argued that the price offered for wood is too low to cover the unexpectedly high cost of harvesting.

Advantages and scope of farm forestry

Supporters argue that commercial farm forestry has a positive impact on rural development. It can stimulate economic activity in rural areas without importing skills or capital; it is a form of cash-crop farming that does not entail dependence on export markets and their fluctuations. It can reduce the need to import paper, pulpwood or other wood-based products. It can relieve critical shortages of poles and increase the supply of fuelwood.

Promoting agencies like farm forestry projects because the profit motive

F. Botts/FAO

Food can be grown between the rows of trees during the early years after planting. A Harijan farmer ploughing in a six-month old eucalyptus plantation in Gujarat, India.

---

attracts farmers into tree growing and then keeps them enthusiastic. The agencies have often found they can simplify programme design, reduce running costs, and give more autonomy to the farmers. The social complexities that have hampered development of communal woodlots are generally avoided.

The direct investments required are relatively small, and any loans or incentives provided to farmers are potentially recoverable when the wood is sold. If the tree growing is truly economically viable, the programme can become self-sustaining. The need for demonstration plots, free seedlings, subsidised credit and other promotional measures will diminish as the financial merits of tree growing are revealed.

The scope for farm forestry is obviously largest in countries with a high commercial wood demand, but fuelwood is seldom the most profitable product for a tree farmer. Industrial demands for lumber, woodfuels, and raw materials for pulp and rayon mills and other industries can create a substantial market for wood.

Many parts of Africa lack both large cities and large wood-using industries.

Many of the existing wood markets are supplied from sources where the wood is collected free of charge. Unless public policies can force industrial and urban consumers to begin paying more for wood, tree growing may not be profitable at all.

Criticisms of farm forestry

Despite their proven ability to get trees grown, some farm forestry programmes, particularly in India, have come under considerable criticism. Critics make three main kinds of charges: that farm forestry benefits the rich at the expense of the poor; that it fails to deliver the social and environmental benefits frequently promised; and that it may actively harm the poorest in the community. Each of these criticisms is sometimes justified.

Although any farmer with some land to spare can plant trees for profit, richer farmers are usually in the best position to take advantage of farm forestry. Because their landholdings are bigger, they are able to allocate more land and labour to tree growing. They can also afford to take greater risks and to wait longer for the benefits. While many small-scale farmers have participated in India's farm forestry schemes, there is no doubt that larger farmers have made the most profits.

Programmes that provide subsidies to all tree farmers are therefore open to the accusation that they are simply helping the rich get richer. In Gujarat, for example, the current limit on free seedlings supplied in polythene pots is 10,000. At a cost of Rs 0.20 ($0.02) each, this amounts to a subsidy of Rs 2,000 (about $200) for those farmers whose landholdings are big enough to enable them to take the full number. This is a considerable sum in an area where the daily wage rate for an agricultural labourer is only Rs 10 ($1.00).

Because of this criticism, free seedlings are to be phased out of the Gujarat programme from 1984. In other programmes, the limit has been set lower so that large farmers do not get a disproportionate share of the free seedlings on offer. If they want to plant more than the limit, they must buy them.

One of the social benefits usually claimed for farm forestry is that it will increase the local availability of fuelwood. But even when a programme is successful in getting a large number of trees grown, it will not necessarily result in a local increase in the availability of wood for fuel.

Farmers will naturally sell wood to the highest bidder. If there is a good market for building poles or pulpwood, then wood may be too valuable to burn. And since the most lucrative market will often be in the cities, the bulk of the wood may end up leaving the villages, with the result that the only addition to local supplies will be from the waste and trimmings obtained when the trees are prepared for sale.

Even where the wood is sold locally, it will only go to those who can afford it. This will often rule out the landless and the poor, who will have to continue to seek free fuel. The depletion of common forest resources will therefore continue. Those burning dung because they cannot afford to buy firewood will not be directly helped.

Farm forestry also offers no guarantee of improving the environment by preventing erosion and protecting watersheds. Farmers planting for the market will do so in a manner that will bring the highest, most secure rate of return. While tree planting on degraded hillsides and other barren lands might be desirable from a national or community perspective, farmers may choose to grow trees on their most fertile land.

This does not mean that no environmental benefits can come from farm forestry, or that farmers will never plant trees in areas that need them for environmental protection. But commercial tree growing is unlikely to be concentrated in the places at greatest environmental risk. In many cases, the most threatened areas are not even privately owned, ruling out farm forestry as a means of protecting them.

For landless and poor farmers who cannot participate, farm forestry may cause a variety of problems and even aggravate those it is supposed to solve. It may actually reduce the local availability of fuel and fodder despite the increase in total wood production. This can happen where traditional custom permits farm labourers and other poor villagers to graze animals on farmers' fields after the harvest and to gather dung and crop residues for use as fuel. If these lands are converted to tree growing without an alternative being provided, the poor will be deprived of an important free good.

Farm forestry can also harm the poor by reducing local employment opportunities, since tree crops usually require less labour than crops such as cotton or rice. One of the reasons why wealthy farmers are turning to farm forestry in parts of India is precisely because it reduces both their labour costs and the problems of farm management.

Finally, many observers worry that the large-scale conversion of good cropland to trees will cut into food production, raising local food prices to the detriment of the poor. Any non-food crop, of course, can have the same impact, and a share of current farm forestry is on lands that formerly grew cotton or other cash crops, not food.

Defenders of farm forestry point out that food can be imported whereas firewood cannot. The strength of this defence depends on the degrees to which programmes actually do increase fuelwood supplies and do help improve the incomes of the poor — the food consumers at greatest risk. In any case, trees are probably not yet serious competitors for land in any large region; foresters in Gujarat claim that so far less than 1% of the state's agricultural land has been devoted to farm forestry. But certainly this is a potential problem that warrants serious consideration by planners.

In India, arguments over farm forestry have been acrimonious. A major 1981 study by the Indian Institute of Management (IIM) in Bangalore was scathing in its criticisms of a World Bank-supported farm forestry project in the Kolar district of Karnataka State. The report claimed that, rather than going for fuelwood, 80% of the wood being grown was feeding a nearby viscose fibre mill, and the rest pulp mills. Though the figures have been hotly disputed, the study also asserted that by 1987, some 200 million man-days of labour could be lost each year as a result of switching to tree planting. It concluded:

> "...in spite of commitments to the satisfaction of the basic needs of the deprived rural population, none of the objectives of the programme seem to have been satisfied. The failure, in our view, arises from simplistic assumptions about the production and distribution of primary products. One simplistic assumption is that just growing more trees will satisfy the basic needs....
>
> The second assumption... is that merely producing more of any commodity in a particular locality automatically amounts to a higher availability of that commodity to the local people. This assumption fails under conditions where the purchasing power of distant urban-industrial groups can pull out resources without allowing them to trickle to local people whose need is greater, but who have very little purchasing power."

The IIM study has been widely discussed both in India and abroad. This and other concerns have led the Karnataka government to put a limit on eucalyptus planting in the state.

Some of farm forestry's problems arise because its supporters imply that growing trees for private profit will automatically bring social and environmental benefits to the local community. A failure to provide such benefits is not, after all, peculiar to farm forestry; it applies equally to many other cash crops that are grown without objection.

But critics also say the problems go deeper than the definition of goals. They charge that in some projects the World Bank and the Indian government are spending public funds to help the well-off become richer and the poor and landless become poorer. Money spent in the name of 'rural development' could actually alter rural social patterns for the worse, creating an area of large tree farms and driving the poor, who once earned money working in annual food harvests, into the cities.

There are few reliable data to substantiate or refute the claims made on either side about the ultimate social impacts of farm forestry. In practice, the effects are sure to vary considerably depending on local conditions. The more programmes can be designed to involve smaller farmers, the wider the benefits will be spread. And if commercial farm forestry is accompanied by other forms of community forestry specially designed to aid the poor, some negative effects can be offset.

Many complex questions must be answered to assess the social effects of farm forestry. How big is the market for wood and when will it be saturated? Are farmers planting trees instead of crops (in which case direct farm employment may fall) or trees plus crops (in which case it may rise)? How much employment is being created 'downstream' in wood-using industries? How does the increase in wood output affect the cost of fuelwood of timber and poles for housing? What is the effect of land conversions on local or national food prices?

Perhaps one of the greatest dangers of commercial farm forestry lies in its very success. Because it can bring such rapid results, governments and aid agencies can be lulled into thinking that the fuelwood problem is about to be solved when in fact the poor are being largely left out. Programmes to help landless and marginal farmers grow trees tend to be slow-moving and difficult to manage in comparison with entrepreneurial private planting. The temptation to focus governmental support on those aspects of community forestry that show the quickest, most dramatic results must be resisted.

Eucalyptus: friend or foe?

Widely used in planting programmes, eucalypts (trees of the *Eucalyptus* genus) comprise one of the best known and most versatile groups of tree species. There are roughly 500 known species, adapted to a broad range of soil and climatic conditions. Most of the species are native to Australia, but eucalypts have been introduced all over the world — some at high altitudes where few other trees can survive, some in near-deserts.

Many are hardy and fast growing. The strong, dense wood provides good timber and poles and is an excellent raw material for making paper and rayon. In Brazil, massive eucalyptus plantations supply charcoal for the steel industry. Many eucalyptus species grow fast under difficult conditions, coppice vigorously and resist fire damage. The young trees are also unpalatable to livestock, which greatly reduces the problem of protecting them from damage.

The extensive use of eucalyptus trees in planting programmes has, however, come under vehement attack. Several charges are made: that they deplete soil nutrients; that they lower the water table because of their heavy water consumption; that because they are often grown as 'monocultures' — tracts of a single species — they are vulnerable to a sudden wipe-out by a new pest or disease; they produce 'ecological deserts', with little fauna, notably birds; and that they fail to provide fuel and fodder, the products most needed by the poor. Some ecologists also worry that chemicals from the fallen leaves and bark of certain species may render the ground unhospitable for other crops in the future.

Eucalyptus has been defended with equal enthusiasm by its supporters. With respect to nutrients and water, foresters agree that fast-growing eucalyptus trees use a lot of both. But they point out that this is true for all fast-growing tree species. When trees are grown in close proximity to crops, there is bound to be a trade-off between tree productivity and crop growth. But if the trees are producing something that farmers need, then they may well consider the trade worthwhile.

The exact effect on water tables depends on many factors, such as the properties of the soil and what plants the eucalyptus is replacing. If the previous crop was a water-hungry one such as rice, the water table could rise. The fact that tree roots break up the soil structure and improve water percolation following rainstorms will also tend to raise the water table. Where there is a subterranean impervious pan, the roots of some eucalypts may break such pan and allow the downwards drainage of swampy areas. If the trees are replacing slow-growing scrub, on the other hand, the water table might fall. The effects can only be predicted through a careful site survey.

As for the monoculture argument, eucalyptus supporters reply that it could logically be applied, with even greater force, to most major food crops. Though foresters agree that there are risks with all monocultures, they argue that this has not prevented coconuts, rubber and other tree species from being grown successfully as monocrops for years. Many of these, because of

G. Foley/Earthscan

Eucalyptus poles grown under a farm forestry project in Gujarat are prepared and graded ready for sale.

selective breeding, have a much lower degree of genetic diversity than a typical eucalyptus plantation.

Dr Gerardo Budowski of the research centre CATIE (Centro Agronomico Tropical de Investigacion y Ensenanza) in Costa Rica commented on the question of ecological diversity:

> "As to ecological deserts, it all depends with what a eucalypt plantation is compared. If it is with a nearby natural, mixed forest, there is no doubt that the latter is much richer in fauna; but if such a plantation is compared with a nearby scarcely covered slope, as for instance a burnt over savanna, then it is highly probable that there is more animal life, including nesting birds, in the eucalypt stand."

The charge that eucalyptus does not yield the products local people need is answered in part by the farmers themselves. In Gujarat and Karnataka they are choosing to plant eucalyptus precisely because it is the most useful and profitable tree for them. The wood is used primarily for poles and pulp not because it does not burn well (though in some areas people say they don't like the smell of its smoke), but because it is too valuable to burn. Budowski adds:

> "In the highlands (altiplano) and interAndean valleys of Bolivia, Peru, Ecuador and parts of Colombia, eucalypts are by far the species most sought for fuelwood, often in combination with other uses such as mine props, poles, and timber. In fact millions of inhabitants of these regions, including a large proportion of Indians, depend exclusively on eucalypts for their fuelwood supply in this region."

Obviously, although prunings provide some fuel, monocultures grown for poles or other commercial markets leave the fuelwood and fodder problems of the poor unsolved. But this is an intrinsic problem with commercial farm forestry, not an evil unique to eucalyptus. If community forestry programmes are intended to meet local fuel and fodder needs directly, then species will have to be chosen with those ends in mind. And if growing these species is not profitable enough to spur great farmer interest, subsidies or other support measures may be necessary.

Another reason eucalyptus is so widely used is simply that foresters — and peasants — are so familiar with it. In recent years, plant scientists have identified many other species that are fast-growing and well suited to one goal or another. Over time, current research efforts should also result in the genetic improvement of many species. Use of a broader array of species in planting programmes will have ecological advantages. The question of whether intensive eucalyptus cultivation damages the land's future capacity to sustain other crops requires urgent research attention, since reconversions of some lands to crops may well be desired in the future.

Tree growing for family use

Some programmes aim to increase tree growing by farm families primarily to meet their own needs. These tend to be pursued in areas where commercial tree growing is limited by the lack of an adequate market, or where farmers' small landholdings and need to grow their own food limits the area that can be given over to trees. The line between commercial and family farm forestry, however, is seldom distinct; families may sell a portion of their tree products and use the rest.

The distribution of tree seedlings forms the backbone of most programmes of this kind. People can usually get seedlings from the wild if they want them. But if nurseries are established a wider range of species can be offered, along with improved varieties that would not otherwise be available.

The potential benefits are many. Making farmers more self-sufficient in fuelwood both guarantees their own supplies and reduces the pressure on local forests. This in turn increases the supply of products from common forest lands available to the poor and landless, who do not have the opportunity of growing trees themselves.

People who engage in small-scale forestry are usually interested in producing a wide range of products: these may include poles, fruit, nuts, edible leaves and shoots, animal fodder, tannins, dyes, bark, fibres, medicines and various gums and oils. People may also want to grow trees for environmental or aesthetic reasons. Though fuelwood may not be the principal reason why people plant trees, it is usually produced as a useful by-product.

Scope and limitations

With no market to dictate what types of trees should be planted and in what quantities, programmes must be carefully attuned to farmers' priorities. Planners from outside the area must avoid trying to impose their own preferences on local people. What an outsider may regard as an extremely burdensome search for fuelwood may be regarded by locals as quite normal, or as a far less pressing concern than something else. If people do not feel firewood scarcity is a high priority problem, they are unlikely to go to a lot of trouble to solve it.

A recent survey in Malawi found that though people said they were aware that fuelwood was becoming scarcer, they were most concerned with the shortage of building poles. As a result, they were more interested in planting trees for poles. In the hill areas of Nepal, a survey disclosed that the overriding need was for tree fodder for buffaloes.

Some plantation programmes in Senegal have shown such poor results that planners assumed there was a local objection to tree growing, or at least a lack of awareness of its benefits. They discovered, however, that the reluctance to plant trees for fuel did not extend to the planting of other types of trees. A report by US forestry expert Fred Weber after the first year of a village plantation project included the following observation:

> "Our visits to village sites conclusively showed that as far as the villagers are concerned, neither firewood nor woodlots for individual villages are foremost in their minds. Rather, what is important to the local population are points not covered in the project documents: shade trees in family compounds, food or fruit trees, small clumps of trees in unused corners, public places, etc. Since these kinds of tree planting efforts can make just as valuable a contribution, not only in wood production but also in socio-economic as well as ecological terms, why not shift project accents in this direction — particularly if that is what people prefer?"

Frequently, the fact that trees provide multiple benefits is what makes them most attractive. Some of the most popular traditional species such as the baobab and the neem tree have dozens of uses, and play a very important role in both the rural ecology and the local subsistence economy.

In Costa Rica, farmers surveyed in 1983 said they were prepared to plant trees for a host of non-financial reasons, including shade for crops, wind protection, soil conservation and simply making farms look better. In particular, farmers preferred species with multiple functions.

Tree growing for family uses is exempt from the more serious criticisms charged against farm forestry. Because it does not generate any income directly, planting will usually be restricted to farm boundaries and unused land around the farm or home. There is rarely any question of converting croplands to tree growing or of reducing the local availability of wood. Nor is there an adverse effect on employment. The only serious question is whether programmes are able to produce results commensurate with the public funds spent on them.

Chapter 7

Helping people grow trees: communal forestry

Communal forestry programmes involve growing trees on public or communal land as opposed to private farms. The theoretical attractions are considerable. Getting communities to work together is often the only way to halt the degradation of communal and public lands. Such projects can allow landless people to take part in forestry activities and obtain benefits otherwise reserved for landowners.

Two countries, China and South Korea, have undertaken communal forestry on a massive scale. The Chinese initiated communal tree planting in the mid-1950s. Backed by the party propaganda machine, the effort was carried out through the newly established system of communes, production brigades, and production teams — in which work was usually obligatory.

The total area planted was reported to have reached 28 million hectares (70 million acres) by 1978. In addition, five billion trees are said to have been planted along roads, rivers and canals and around houses and villages. The reliability of these official estimates is in great doubt, and the survival of planted trees was quite low in many areas due to a lack of proper technical control and follow-up. However, even with widespread failure (and huge continuing needs for reforestation), China has apparently pursued communal forestry on an unusually large scale and with considerable success.

Although it has a different political system from China, South Korea, too, has combined central governmental direction with local community participation to achieve reforestation. Since the land reform legislation of 1949, almost three-quarters of South Korea's forest lands have been in private ownership, generally in smallholdings of 2-3 hectares (5-7.5 acres). By the 1960s, extensive deforestation in this hilly country was having severe environmental effects and resulted in a shortage of woodfuel for household heating in particular.

The government used its New Community Movement, launched in 1970 to mobilise villagers for rural development, to address the forest crisis. An ambitious 10-year National Forest Plan was launched in 1973 and implemented through a network of Village Forestry Associations. Sites for planting were selected at a village level; profit-sharing arrangements between landowners and the Forestry Associations were laid down by the government.

The achievements surprised even the planners. By 1977, five years ahead of schedule, the overall planting targets had been met, and in some cases exceeded. In all, over one million hectares (2.47 million acres) were planted,

Peyton Johnson/FAO

Tree seedlings being tended at a community forestry nursery in Thailand.

an extraordinary achievement for one of the world's most densely populated countries.

Both China and South Korea combine unusually strong central authority with long traditions of community cooperation. While their successes (and failures) in communal forestry offer many lessons for planners elsewhere, neither offers a simple model for wider application.

Ethiopia is perhaps the only African nation to develop a large-scale communal forestry programme. The revolutionary government which took power in 1974 has organised the farmers into more than 28,000 Peasant Associations with over seven million members. The Ministry of Agriculture has organised these into efficient tree-planting units, especially in the highland regions north of Addis Ababa which have ancient traditions of communal work.

Kabede Tatu, head of the Ministry's Soil and Water Conservation Service, said in 1984 that over the previous seven years some five million seedlings had been planted and 700,000 km (435,000 miles) of farmland and hillside terracing have been dug. Some 3,500 peasants are trained each year in the basics of reforestation and soil conservation, he said. The total responsibility for the fieldwork is left to the Peasant Associations, which are now able to mobilise 30 million man-days of work per year, according to Kabede. Workers are paid through UN World Food Programme 'Food for Work' schemes. There is little data on survival rates, and officials freely admit that many trees have died in the three years of drought affecting the highlands.

The peasants of Wollo Region, where impressive reforestation and terracing can be seen along the entire length of the region's main north-south road, appear to be motivated as much by desperation as by food payments and government encouragement. Almost all the hillsides in this rugged landscape have been overgrazed and overfarmed, so that any rains bring many upland fields sliding down onto the fields in the valley. The peasants say they are now convinced they must stabilise these slopes to continue farming in the region.

Collaborating with local communities

Forest services often use their own resources to establish plantations on public or communal land. These can include officially designated village common lands, roadsides, canal banks and railway boundaries as well as areas under the control of public works departments, irrigation authorities and other state institutions.

In these programmes, the forest department usually pays for planting and bears the costs of seedlings, fertilisers and other necessities. Labour is usually hired locally; this creation of employment is often cited as an important benefit of such schemes.

Planting on community land requires prior consultation with the local bodies concerned. In many countries, people are suspicious of any government interference. They may be particularly concerned that the government will later appropriate community lands on which new forests have been established.

Even when the land is publicly owned, agreement with the local community on using it for tree growing is important. Though local people may have no official title to the land, they frequently have customary rights to use it for grazing, fuel collection or other purposes. Resolving this issue is crucial; no matter how much money is spent on fences and guards, plantations cannot be protected against a hostile public. Many attempts to force communal forestry programmes on people have ended when the seedlings were pulled up, burned or eaten by livestock.

Converting scarce grazing grounds to plantations can pose special hardships for villagers. To help overcome them, planting is sometimes staggered over several years so that the area of restricted access is kept as low as possible at any one time. As the trees get big enough to withstand grazing, the area can be opened to livestock while another is being planted.

People may also be given access to young plantations to cut grass fodder by hand. Often the amount of herbage so produced is actually greater than before, especially when improved grass species are planted between seedlings.

In most cases, the local community's main incentive to cooperate is the prospect of financial returns. In most Indian collaborative schemes, the wood produced is to be auctioned and the revenues divided between the forest service and panchayat (village council) according to terms agreed upon in advance.

This can be an effective way of generating community funds. In Tamil Nadu, some of the larger plantations on the foreshores of communal irrigation reservoirs, cover more than 50 hectares (125 acres). At current market prices for firewood, these are expected to earn the equivalent of approximately $500 per hectare ($200 per acre) after 10 years.

Even after the forest department deducts the plantation establishment costs and takes 50% of the net revenue, the community budget receives a helpful input. Panchayats have complete control over such funds, and many intend to use such income for schools, health facilities and water supplies. In the 1982 and 1983 planting seasons, a total of 42,000 hectares (104,000 acres) of community plantations were established using this approach.

A similar scheme has been running in Gujarat since 1974. With the agreement of local panchayats, the forest department plants 4-5 hectare (10-12 acre) plots on communal land. Villagers are allowed onto the plots to cut fodder during the growing period, and are promised a 50% share in the net profits when the trees are sold. The scheme was slow to get under way, but interest began to pick up in the late 1970s. By 1982, woodlots covering a combined total area of 28,000 hectares (69,000 acres) had been established in 4,000 villages, though low seedling survival rates have been reported from some.

Beyond informal assistance in the protection of plantations from livestock and wood poachers, the role of the community in such programmes is largely passive. But some of the agencies promoting the schemes see them as only the first step towards a more genuine acceptance of responsibility by the local community.

In both Tamil Nadu and Gujarat, the forest departments want to transfer responsibility for plantations to the villages. But many of the communities involved in the present schemes prefer the system as it is. Many villagers feel that allowing an outside agency to organise planting and pay for guards has considerable advantages, helping to avoid disputes about the distribution of work and the allocation of responsibility if something goes wrong.

Increasing community involvement

Other programmes endeavour to increase the degree of local participation and control. They normally use either village-owned common lands or state lands that have been specially designated for community control. Though the forest department often provides assistance of various kinds, the community is expected to assume most of the responsibility for planting and looking after the trees; it also gets most of the benefits. In theory, locally controlled projects can be better attuned to local capabilities and needs, and are cheaper to run.

The main task of agencies promoting such programmes is to stimulate and organise community activity. This may be done through publicity efforts, convening village meetings and providing whatever technical assistance is needed. Sometimes a local person is appointed or hired to act as a motivator to help generate awareness and cooperation within the village.

Subsidies may be provided in the form of free seedlings or fertiliser. But the main inputs into the programme come from local resources. Labour is supplied voluntarily or workers are paid out of community funds. The community takes responsibility for protecting the plantation, harvesting it and distributing the benefits.

Since they rely so heavily on local initiative, cooperation and resources, participatory programmes run into major problems when any one of these ingredients is missing. Getting them started at all is often difficult.

In Tanzania in the late 1960s, the government planned that every village would start a woodlot to supply its own fuelwood needs and improve the physical environment. Up to 1982, the success rate was abysmal, though recent reports are more optimistic. Villagers are sometimes unhappy with the arrangements for distributing the income from the sale of trees. Like farmers everywhere, Tanzanian villagers are suspicious of any financial dealings undertaken on their behalf, and some have feared that members of the village government might embezzle the money. In other villages, factions were suspicious that their rivals might gain an unfair share of the proceeds.

Progress with village woodlots in the Sahel has been similarly disappointing. Since the 1968-73 drought, enormous efforts have been made to promote community forestry there. Between 1975 and 1982, more than $160 million was spent on various community forestry projects.

According to a 1982 report prepared for the US Agency for International Development, the achievements to date amounted to "little over 20,000 hectares (50,000 acres) of plantations that do not grow very well". Most of the few successful village woodlots were planted and managed entirely by the forest department, with minimal local involvement.

Here too, villagers are suspicious about where the profits are going. According to US sociologist Marilyn Hoskins, when peasants in one area were asked whom they thought would obtain the return from the plantation, their answers included the forestry service, the village chief and the project's foreign designer. Some did not know and were unwilling to guess; few believed they personally would receive any of the wood when it was harvested.

The few successful projects give some hope and guidance for the future. One is the windbreak planting in the Majia Valley in Niger. This has been carried out at a low cost and with enthusiastic local support. As a result, some 20 km (12 miles) of windbreaks per year have been planted since 1975. In the same region there have been successes in dune stabilisation with trees and in efforts to encourage the planting of gum and fruit trees and live fences.

In India, participatory village woodlots have been tried in Gujarat and Uttar Pradesh. The forest department supplies technical assistance and free seedlings, but all other planting and protection costs must be borne by the village. On harvesting, all the proceeds go to the village.

The response has been poor in both states, with tree planting well below target. Reasons given include the lack of common interests within villages, a mistrust of the profit-distribution system, disputes over land availability, and the shortage of panchayat funds to pay for plantation establishment.

Both states are now emphasising programmes in which the initial inputs are provided by the forest department. But the departments hope they can eventually shift a greater degree of responsibility to the local community. A similar gradual approach to the development of participatory programmes is being followed in Tamil Nadu, Jammu and Kashmir and Harayana.

Scope for communal forestry

Communal forestry is considerably more demanding than farm forestry. Where tree growing conditions are difficult or local organisations are weak, it is much harder to start communal programmes than those in which farmers participate for direct personal profit.

In few if any villages, whatever the country, do all the residents have identical interests and priorities. Communal programmes must be based on compromises on matters of deep concern to all. Rigid social stratification, such as in India's caste system, make any communal action extremely difficult. Thus communal programmes require careful preparation and negotiation.

Communal programmes also presuppose the existence of competent and representative local organisations that can plan and implement projects and distribute benefits fairly. But often villages have no organisation with such capabilities. Where it does not already exist, creating a body just to manage community forestry is difficult. Some of the more successful village forestry efforts, as in South Korea and Nepal, have built upon traditional village institutions for forest management.

Existing village organisations are often fiercely independent. Though promoting agencies may have preferences about how the programme should be run, they may be able to issue only guidelines rather than enforceable directives. When negotiating the terms of programmes with village organisations, it is up to the programme promoters to ensure that the interests of the poor and of women are protected, as they are rarely represented adequately in existing bodies.

Despite the problems inherent in communal forestry, it can work when conditions are right. Although the communal forestry achievements of some of the Indian states have been dwarfed by those of farm forestry, they represent substantial achievements in themselves. These programmes have shown that planting with the collaboration of nearby people along roads, canals, railways and other wastelands is feasible and beneficial. But target areas must be carefully selected, the programme promoters must be patient, and no attempt must be made to force tree growing upon people.

Recent experience in Nepal has shown that communal forestry projects there have considerable promise. As the resident chief technical advisor to the programme, E. Pelinck, states:

> "...we have found that communities are generally quite interested and their participation in afforestation and forest management is the basis for community forestry in Nepal. After three years of operation there is no indication that this has been a wrong assumption. During a socio-economic survey it was found that 85% of the people are prepared to make common grazing lands available for community plantations. Survival counts of plantations established in 1981 and 1982 indicate that the main reasons for failure, where it occurs, are technical or administrative rather than socio-economic."

In the past, many planners burdened communal forestry with naive optimism and unrealistic objectives. Not surprisingly, many projects failed to live up to expectations and this led to some disillusionment. But now that the difficulties and requirements of communal forestry are becoming clearer, more realistic programmes are being devised.

The main lesson from the past is not that communal programmes should be abandoned; rather, their design and implementation should be improved. Village woodlots cannot provide a magic solution to the firewood crisis, but communal forestry efforts of various kinds can contribute to the betterment of rural life.

Forestry for the landless

Both to aid poor, landless families and to get forests established, several governments have offered people temporary or permanent rights to public lands. The idea is to turn landless people into tree farmers by giving them security of tenure and other incentives.

Under a new scheme in West Bengal, landless people are given official title to small plots of government-owned wasteland. Seedlings and fertiliser are provided for free. In the second and third years, a small cash incentive is paid on the basis of the number of surviving trees. This cash is provided because of the extreme poverty of the people involved; daily wage rates in the area are the equivalent of 40 US cents. Receiving even a small cash return from the trees in the first couple of years encourages people to persist with caring for the trees until they are ready for harvest, in eight years, when the planters receive the full proceeds from their sale.

A similar land-allocation approach is being used in Gujarat to reforest degraded areas of forest reserves. Tribal families who through their former practice of shifting cultivation were helping destroy the forest are paid to plant and henceforth protect a new plot each year. Participants are promised 20% of the profits when their plots are harvested after 15 years. Thus they have a strong financial stake in tending the allocated lands well.

A scheme in the Philippines is encouraging farmers who were previously squatters on government land to grow trees to supply wood-fired power stations. The farmers do not receive formal title to the land, but by joining in the programme they gain access to credit facilities, technical advice and a guaranteed market for their wood. All benefits and management transactions are channelled through tree farming associations, consisting of groups of 10-15 families.

The first of 14 power plants currently under construction was opened at Cape Bolinao in Pagasinan Province in February 1984. Designed to generate 3 Megawatts of power, it is located in a 1,100-hectare (2,700-acre) tree plantation that provides employment for about 120 tree-growing families. Power plants are nearing completion in several other parts of the country, but tree growth in some of the plantations has reportedly been much less than expected. Some power plants may have to buy wood from elsewhere until the trees catch up.

Land allocation forestry schemes also involve a risk that the land will be diverted to crop-growing or that people will even assert a permanent right of occupation. Once poor families start farming the land, it is often politically difficult to force them to leave — and thus the forest may permanently shrink.

If they are to be more than simple wage-labour operations, land allocation schemes must provide people with a clear economic incentive to raise trees to maturity. They are therefore restricted to areas where there is a strong commercial wood market. But where the land is available and the conditions are right, such schemes offer a chance to bring non-productive land into use while giving some of the poorest members of the community a chance to build assets in the form of growing trees.

Taungya systems

The 'taungya' system is a well-tried means of giving people temporary farming rights in return for their work in establishing forest plantations. It was first used in Burma in the 1850s as a means of replanting teak forests and has since been copied elsewhere. In the original version, people were allowed to plant food crops between the young trees as payment for their tree-planting work. The labourers had no rights to permanent settlement.

A variant of the taungya system was introduced in Java around 1856 and is still extensively used there. Normally, families are allocated 0.25 hectares (0.6 acres) each and are permitted to grow food crops for two years. Crops are interplanted between alternating rows of direct-seeded teak and *Leucaena leucocephala,* which acts as a cover crop.

The taungya method can be considerably cheaper than other means of establishing plantations. Though it is more labour-intensive than conventional planting methods, only part of the costs have to be borne by the forest service or plantation owner. However, after the first couple of years the trees and food crops are in competition, meaning inevitable conflict between the interests of the cultivators and the forest owners. In some cases farmers have deliberately damaged trees to prolong the period of crop cultivation.

Taungya plantations tend to be successful only in areas of acute land hunger and underemployment — where families have few alternatives. They have been criticised on the grounds that they exploit the weaknesses of marginal and landless farmers. A report by India's former Inspector General of Forests, S.K. Seth, drew the following uncompromising conclusions:

> "The (taungya) system is frankly exploitative in concept and operation and cashes in upon the needs of the landless and poor people to serve its own ends. The much vaunted incentives are only a cloak for uninhibited exploitation, as the savings effected by the Forest Department are many times more than the expenditure incurred on elementary conveniences provided to the working force."

Not all taungya systems need be subject to such criticism. Workers can be given a guaranteed wage for their labour in addition to the crops they are able to grow. They can be supplied with welfare facilities and opportunities for earning money by other means. This approach has been used both in Java and in the 'forest village' programme in Thailand.

Thus taungya can provide considerable benefits to the families engaged in it, and can be an important part of an integrated approach to rural development. But these benefits do not accrue automatically. Unless programmes are specifically designed to address the needs and aspirations of those taking part in them, they can easily become little more than a means of exploiting their poverty.

Management of natural forests

Recently, many foresters have begun to argue that involving communities in the improved management of existing forests can be an effective way to supplement or even replace plantations. This interest has been stimulated by the high costs and low returns of some plantations and also by the growing recognition of the many benefits rural people derive from diverse natural woodlands.

Even countries that have suffered severe deforestation still have forest reserves, though these are often degraded as a result of poaching and illegal grazing. Many villages are surrounded by degraded patches of forest under national or common village ownership. In semi-arid zones such as the Sahel, large common lands outside the forest reserves have scattered tree cover.

In most cases these natural woodlands, however degraded, play key roles for the poorest of the poor: the small-scale farmers or landless labourers who have little access to trees or other biomass from their own land. In the Sahel, the natural woodlands provide people with wood and a whole range of other essential products.

Because they are publicly owned yet accessible, the natural forests suffer the effects of the 'tragedy of the commons'. The individual wood collector or herder, lacking secure long-term rights to the land or its produce, has little interest in seeing to it that the forest's yield remains sustainable.

Sometimes traditional systems of community forest management have been destroyed by well-intending but misguided government policies. Nepal's 1957 nationalisation of forests — including many that had been under cooperative village control or were well-managed by private individuals "hastened the process of forest depletion", says P.K. Manandhar, former chief of Nepal's community forestry programme. Traditional constraints on overcutting were undermined and "people tended to overexploit forest resources which they no longer felt were their own".

Recognising the error, Nepal passed new laws in 1978 permitting nearby forest lands and new plantations to return to village ownership, provided the village agrees to a sound management plan. Since then, where degraded forests have reverted to village control and been well protected, trees have regrown even faster than foresters had predicted. In some cases, protection has been supplemented by the planting of desired tree and grass species. Handcutting of grass, combined with stall feeding of livestock, has replaced the former destructive grazing, and resulted in more total fodder that had previously been chewed off these lands. More fodder in turn means more milk and more dung for compost.

In the Sahel, wood production from new plantations — many of them established by bulldozing existing natural woodland — has often been well below the projections used in justifying the investments. Some foresters now observe that better control of cutting, burning and grazing in natural woodlands, perhaps in combination with replanting, could raise their wood output at a fraction of the cost of new plantations. Moreover, a rich, varied ecosystem would be preserved.

According to one estimate, an increase in the yield of Upper Volta's natural forests by 25% through improved management would produce extra wood equivalent to the output of 500,000 hectares (1,235,000 acres) of plantations. In fact, many plantations in the Sahel have performed so poorly that their yield is only slightly higher than that of unmanaged forests. In some cases, monocultures established at costs of at least $1,000 per hectare ($405/acre) have provided little more wood than the former useless brush that was bulldozed to make way for them.

Attractive in theory, natural forest management faces serious technical and social obstacles. There has been little research on the species and dynamics of the native woodlands of tropical semi-arid zones. Little is known about feasible rotation patterns and how best to incorporate controlled grazing and woodcutting into management.

The social unknowns may be even greater. Forest management requires close cooperation both among villagers and between villagers and foresters. Where an organised, single village can control a patch of forest, as is being done in Nepal, the problems of coordination are reduced. But gaining the cooperation of scattered communities over huge areas, as is needed in parts of Africa, is a tall order. In either case, forestry departments have to learn to work with people, not order them around as they often have in the past.

Management of existing forests has received much less attention than it deserves. It is less glamorous than large tree planting programmes. It offers no prospect of instant success through the exploitation of new miracle species of trees. But as experience in Nepal and elsewhere is beginning to show, it can be effective. Natural forest management could help meet fuel needs in many places.

Chapter 8

Making community forestry work

One of the key lessons of the last decade's community forestry activities is that helping people meet their own forest needs requires radical changes in the purposes, practices and personnel of forestry departments. Traditionally, Third World forestry officers have been concerned only with officially designated forest reserves — with protecting them from encroachers and with selling the timber in them. Rural peasants generally have seen foresters as policemen whose role is to deny them access to desperately needed goods. Relations between rural people and forest departments have often been characterised by mutual hostility and suspicion.

Successful community forestry requires that foresters move out of the forests to help people plant trees around their homes, farms and villages. It also requires genuine popular participation in decision-making — where to plant, what species to plant for what purposes, how to harvest and distribute the produce. Without active public support, neither paid guards nor barbed wire can suffice to protect seedlings from grazing animals or wood poachers. Experience has proved time and again that participation is more than a development cliché; it is an absolute necessity if goals are to be met. But working with people rather than policing them is a new role for many foresters.

A major problem in some countries is a long-standing tradition of corruption and abuse of power by forestry officials. In a comparative study of forestry services in the Asia-Pacific region, FAO consultant David Palin stressed that corruption "is a matter of great concern to senior public forestry administration officials and to politicians met in most Asian countries. Judging from the frequency with which they are mentioned, occasions where officers use their official positions for personal gain abound."

Some local forest officials derive a substantial income from the fines they informally levy on villagers obtaining wood from forest lands. Such people are scarcely likely to be suitable or enthusiastic recruits to the seemingly more humble role of extension agent, helping people to plant their own trees as alternative sources of fuelwood and other products.

Forest officers must become educators and community organisers rather than policemen; thus they need training in sociology as well as silviculture. More female foresters who can reach women and involve them in solutions to the fuelwood problem are also essential.

Backed by international aid, many governments in Africa, Asia and Latin America are restructuring and re-educating their forest departments to implement people-oriented forestry. New training schools for junior and mid-level personnel have been established in many countries, and professional-level curricula are being amended to reflect today's imperatives.

In some countries, a new generation of foresters with more relevant, people-oriented training is now being created — a hopeful sign for the future. While dedicated, technically proficient leadership at the top of forest departments is crucial, this cannot be translated into results on the ground without equally dedicated field workers. The need is for 'barefoot foresters' — men and women who may have less scientific training but who are willing to undertake physically demanding, patient extension work in remote rural areas.

The relations between forest officers and rural people are sometimes so poor that consideration should be given to using other agencies for forestry extension work. In any case, it would often be desirable to incorporate forestry into the work of existing agricultural extension services, which usually have large numbers of people already working in the field. This could not only save funds but also result in greater recognition of the often ignored links between farming and forestry, and of the potentials for agroforestry to help solve both food and fuel problems.

Non-governmental organisations

Even in the best of circumstances, government agencies often have traits that hamper implementation of community forestry: inflexibility, bureaucratic tendencies and difficulty in communicating with people at the grass roots. In several countries, voluntary and non-governmental organisations have proved effective in community forestry programmes, helping motivate people and bridging the gap between local communities and forest services. Their participation can give local people a voice and a way of bypassing obstructive local officials.

Relations between villagers and the forest service had long been poor in the Ranchi area of Bihar state in India. In 1977, a consortium of local voluntary organisations was formed to promote community-level forestry. Immediate liaison was established with the Chief Conservator of Forests, who quashed all cases of violation of forestry regulations pending against the people in the area. Today, villagers are collaborating enthusiastically with the forest department in a number of projects.

The Anandniketan Ashram (religious centre) in south Gujarat has been involved in a variety of village development projects emphasising the very poor over the past 35 years. It launched a tree growing programme in 1981. Within a year, 1.25 million saplings had been planted along embankments and on waste lands, with a reported survival rate of 90%. Tree growers' cooperatives have been established, and there are plans to establish nurseries and raise the annual planting rate to 10 million seedlings.

In Kenya, voluntary organisations such as the National Council of Women of Kenya and the National Christian Council of Kenya have also promoted tree growing in collaboration with the Forest Department. They have helped with publicity campaigns, purchasing seedlings, developing nurseries and raising seedlings for distribution to local people.

Though not strictly 'non-governmental', schools can also play an important forestry extension role. Enthusiastic teachers have a considerable influence over the children they teach; they also tend to have a high standing in their village.

The 3,000 pupils of one large secondary school in India are reported to have planted about 100,000 *leucaena* trees in 1982, of which 80% were doing well. In Tanzania, some primary schools in the Dodoma, Arusha, and Singida regions have managed to establish plantations of up to 10 hectares (25 acres) in previously barren areas.

Women and wood

A failure to include women and their special concerns in planning has been a major weakness of most community forestry projects. The negative consequences can be tangible. An evaluation of Nepal's farm forestry programme revealed cases where seedlings were eaten by livestock because the men who planted them had failed to inform their wives, who tend the animals. The same study found that far fewer women than men came to take free seedlings, and when they did, the seedlings had lower survival rates — indications that extension efforts were not reaching women well.

One practical reason for involving women more fully is that they are becoming the effective family decisionmakers in many parts of the world. As more men are forced to migrate to cities in search of work, more women are left in charge of rural households. It has been estimated that a third of the world's total households are headed by women.

Women often have different attitudes than men about tree growing. As the main gatherers and users of forest products, they may be more aware of emerging scarcities and more receptive to planting efforts. Also, women sometimes have specialised knowledge of local tree and plant species that could be of great value to foresters.

Women's preferences about which species to plant may differ from men's. In the Himalayan district of Chamoli in India, discussions about which trees should be grown in village plantations showed that men preferred income-producing fruit trees; the women, on the other hand, wanted trees to produce fuel and fodder.

Tree planting in Wollo Region, highland Ethiopia. Women are the main gatherers and users of wood and are therefore more aware of shortages and often more sensitive to reafforestation efforts.

If programmes are to help women and take full advantage of their potential contributions, planners must have accurate information on their work and roles in the community. Sometimes this can only be obtained by female researchers. Planners have to guard against prejudging the local situation. Most outsiders assume that women must regard lengthy fuelwood-gathering journeys as an onerous burden. But this is not always the case. In one part of Tanzania, women told researchers that they preferred wood collection to other work, finding it an enjoyable time for socialising.

Project planners must take steps to ensure that local women are well represented in village organisations that are negotiating and implementing community forestry. But formal representation is not always enough, since women tend to be subjected to strong domestic and communal pressures to conform to traditional roles. Often women will not speak out if men are present, and will allow men to speak on their behalf even when they know the men are wrong. In Senegal, men told project workers that women do not plant trees. But in fact the fruit trees around the houses are planted by women.

Having women adequately represented in the professional planning and management of projects can help ensure that women's concerns get a hearing. This can sometimes be achieved by granting a strong consultative role to independent women's organisations. But women professionals can also be given explicit responsibility for considering project impacts on women. Finally, the extreme scarcity of female foresters — both professional-level foresters and extension agents — is costly, and needs to be remedied as soon as possible.

Getting the technical package right

The technical problems involved in community forestry are often regarded as minor compared to the many social and economic constraints. In countries with a favourable climate and an experienced forestry service, this may be true. But elsewhere, the lack of proven technical packages is a real problem. This is particularly true in drier regions where plantations often fail due to inappropriate species choice, incorrect management techniques, drought or disease.

In some cases, trees have survived through several seasons of good rainfall, but because they were not sufficiently drought resistant, all were lost when the inevitable drier year came. Trees have also died after a number of years when the tap roots reached a hard layer they could not penetrate. In some Sahelian programmes, termites have destroyed large numbers of *Acacia albida,* eucalyptus and other seedlings. Tanzanian forestry experts have admitted that the "wrong choice of species to match the climatic and soil conditions" has on various occasions resulted in project failures.

The Philippine programme to fuel power stations with wood has been hamstrung by technical difficulties. Partly this was due to exclusive reliance on one species, the giant ipil-ipil *(Leucaena leucocephala)*. Under favourable conditions it grows extremely fast, but it does not grow well everywhere. One ipil-ipil plantation was completely destroyed by snails in the year it was planted. Others have grown poorly because of acid soils or other unsuitable site conditions. As Dutch forester Hans Heybroek states:

> "...any new plantation is a gamble, but especially on a new site, or in a new region. The decision to plant trees often has to be taken with imperfect knowledge and before the chosen species has proved its health, value, and desirability over a full rotation."

Diversification is one of the main lines of defence for foresters. By planting more than one species, the risks can be spread, though not eliminated. Another safety measure is to emphasise local species that are known to be suited to the project area. Though they may be less glamorous for foresters and may not grow as fast as exotic species, the uncertainties are reduced.

In general, the further tree planting activities diverge from local traditions, the greater the risks. Species trials and tree management tests reduce these risks. But these take time.

There is always pressure to achieve quick results. Unless the technical package is right, though, the project may fail completely, dampening public enthusiasm much more than slow tree growth would. And once people's confidence in forestry has been undermined, it is hard to restore.

Chapter 9

Improved cooking stoves

With fuelwood, as with any form of energy, scarcity can be relieved either by increasing the supply or by reducing the demand. In the mid-1970s, as concern spread about the fuelwood problem, interest surged in the possibility of wood conservation through adoption of more efficient cookstoves.

Open fires and traditional stoves, researchers noted, transfer little of the energy of burning wood into the cooking pot. In theory, stoves can be designed that transfer more energy and radically reduce wood needs, thereby saving family time or money and reducing whatever pressure fuel collection puts on forests. If they incorporate chimneys that take smoke out of the kitchen, new stoves can also cut the nuisance and health hazards of current cooking methods. Excited by these theoretical possibilities, stove enthusiasts have made expansive claims about the potential for stoves to save wood and forests.

Experiments with various stove designs have proliferated, carried out by everyone from teams of physicists to local appropriate technology groups and Peace Corps volunteers in remote villages. Several internationally financed community forestry projects have included efforts to disseminate improved stoves.

As scientific theory has confronted field reality, the results have so far been almost uniformly disappointing. Changing something so deeply ingrained as cooking habits, stove promoters have discovered, is not simple or easy. Traditional cooking methods turn out to have many advantages that outsiders had overlooked, and are not always as inefficient as had been assumed. And inventing new designs that can be built cheaply from local materials and are durable and simple to use has proved quite difficult.

This new era of stove design and promotion is only in its beginning stages. Clearly, however, the grandiose hopes many expressed a few years ago must be scaled down. Improved stoves are likely to play some role in solving the household fuel problem, but are no panacea. Compromises among fuel efficiency, convenience of use, durability, cost and cultural preferences are inevitable.

One thing is certain: families will not change their cooking habits because of abstract concerns about national energy shortages or deforestation. They must see real benefits for themselves.

Three-stone fires like this one in Haiti, go back to prehistoric times. They are portable and easy to regulate; they burn fuel of all types, shapes and sizes, and they provide light, heat and a social focus.

Traditional methods of cooking

Most rural people in the poorer countries cook over open fires, usually set between three or more stones, bricks, inverted pots or mounds of mud that support the cooking pot. This is the traditional 'three-stone fire'. Sometimes, especially in Latin America, this open fire is placed on a raised adobe or mud platform. Metal trivets — horizontal metal rings to which three legs are attached — are used in some countries to hold a single pot over an open fire.

Traditional stoves have evolved in many parts of the developing world. Residents of most African cities use light metal charcoal stoves made by local tinsmiths from scrap metal. In Kenya and other East African cities nearly everyone cooks with a metal charcoal stove called a *jiko*, which appears to have been introduced 50 or 60 years ago. Especially in the French-speaking countries of West Africa, urban residents use a similar metal stove called a *fourneau malgache*.

In India, too, metal stoves are often used for cooking with wood or charcoal. Various ceramic, single-pot wood and charcoal stoves are found in Asia. One of the best known is a double-skin charcoal stove called the *Thai bucket*. It consists of a metal bucket with an inner pottery lining. For added insulation, the space between the inner lining and the outer metal skin is filled with a light material such as ash from burnt rice husks. This stove was developed in Thailand around 1920 and is now used all over the country.

Fixed mud-stoves called *chulas* are also common in Asia. (*Chula* is the Hindi word for a woodburning stove.) These are usually constructed from mud or mud and sand. Straw, dung, wire mesh or pieces of metal are sometimes added as reinforcement. Cement plaster may be used to provide a hard outer surface.

Many mud stoves have two or more potholes. In some, these are positioned symmetrically over the fire chamber, with each pot obtaining roughly the same amount of heat. In others, the potholes are placed one behind the other and the fuel is pushed in from the front, so that one pot cooks hotter than the others. Most mud stoves are built by their owners from local materials. Ceramic linings or old earthenware pots are sometimes added to line the fire chamber and increase the working life of the stove. Old earthenware water vessels are also used for the same purpose.

Mud stoves without ceramic linings are not durable. The heat of the fire causes them to crack and pieces then break off around the firebox opening or at the edges of the potholes under the impact of the pot. A typical life for a mud stove is one or two years, but it can be as short as six months.

Efficiency of traditional cooking

Open fires, particularly when used outside, can be very inefficient in the transfer of heat from fuel to pots. But a few simple measures can greatly improve their fuel efficiency. By putting the stones closely around the fire and tending it carefully, a cook can save a good deal of wood.

Many people erect shields against the wind. Ghanaian women often fill the spaces between the stones with mud, while women in the Transkei cook inside the shelter of low rectangular walls. Or the fire may be built in the lee of a wall.

Peasants know when fuel is scarce and, when it is, they tend to conserve it, carefully controlling the fire and recovering unburnt wood when cooking is finished. In Senegal, visiting experts from the Oregon-based Aprovecho Institute commented: "Economy is second nature, and women will quench the fire with water or bury the embers in sand immediately after the cooking is through."

The efficiency of open fires has become a topic of study and confusion. For years, many researchers have assumed that open fires have an efficiency in the range of 3-8%. ('Efficiency' in this case is basically a measurement of how much of the fuel's energy gets into the pot above it.) But methods of measurement have differed or been scientifically imprecise, and no available estimates are widely applicable to people's actual cooking habits.

Some laboratory studies have shown that open fires can, under the right conditions, be much more efficient. Experiments at Princeton University produced efficiencies of around 20% for open fires. Tests by the Woodburning Stove Group at Eindhoven University (the Netherlands) gave similar results. Using specially dried wood brought even higher efficiencies. But these tests took place under controlled conditions; there were no random draughts, fuel was cut into small pieces and the fire was carefully controlled. Such high efficiency is not possible under normal cooking conditions. What these experiments do indicate, however, is that when people need to economise on fuel, they can do so, within limits, even on an open fire.

The relevance of fuel-efficiency measures is limited when people rely on fires for heating and lighting as well as cooking. In higher altitudes and in winter in many parts of the Third World temperatures can become quite cold, and poorer families get what heat they can from the cooking fire. In some unelectrified homes where people cannot obtain kerosene, the fire provides the only available light at night — and sometimes the only available light for cooking. Unfortunately, many stove designs that save wood also reduce the heat and light gained from the fire.

Advantages of traditional cooking fires and stoves

Although it can be dirty, dangerous and inefficient, the open fire offers some important advantages. It can be built for free, usually with no special materials or tools. It can be located anywhere to take advantage of shelter or shade.

An open fire can burn fuel of any size, shape or type and, where cutting tools are primitive, it may be the only practical way of using large pieces of wood. In many parts of the Third World, few people possess an axe or saw.

The open fire is also relatively easy to control by adding or subtracting fuel. The stones can be adjusted to all shapes and sizes of pots. Many traditional diets are based on heavy paste-like mixtures of flour and water that need long stirring; stones can effectively secure the pot and keep it from sliding about when these are being cooked.

Traditional open fires and stoves are widely used because they get the jobs done and represent an effective compromise between the often conflicting requirements of utility, economy, convenience and general compatibility with the domestic environment. This is why they are hard to beat for practical everyday use, though the new stoves may perform better in laboratory tests.

The problem of smoke

From the few air-quality measurements taken in developing country kitchens, it is clear that the concentration of pollutants is often dangerously high. Studies of traditional cooking in indoor kitchens, directed by the East-West Centre in Honolulu, have shown:

> "that cooks receive a larger total dose than residents of the dirtiest urban environments, and receive a much higher dose than is implied by the World Health Organization's recommended level, or any national public standards. The women cooks are inhaling as much benzo(a)pyrene as if they smoked 20 packs of cigarettes per day."

Medical researchers are finding high rates of various acute and chronic respiratory diseases in places where cooking is done in enclosed spaces. (In warmer regions women often cook outside or in well ventilated shelters, so the smoke densities are lower.) The common cold is ubiquitous in, for example, Nepal's hills, and chronic bronchitis, tuberculosis, and eye diseases are prevalent. Given all the other health hazards present, it is impossible to prove the exact contribution of smoke to these disorders. The health impact of household smoke is a neglected area of research — but health may well prove the strongest reason for promoting smokeless stoves.

Even outdoors, cooking over a smoky fire can be unpleasant and irritating for the eyes. Smoke also dirties clothes, furniture, wall and ceiling coverings and other household possessions. In Papua New Guinea smoke is hampering the introduction of rural electric lighting because it fouls the lamp fittings. Women surveyed in several different stove projects have called the reduction of smoke irritation the most attractive feature of their new stoves.

But smoke can have its uses. It cures and preserves foods; it can keep termites out of thatch roofs and walls. In Central America, cooking fire smoke binds the straw of the roof, making it stronger and more waterproof. People in this area also hang ears of corn above the fire; the smoke eliminates insects, so the corn can be kept for six or seven years.

Appropriate stoves

Designers can readily come up with stoves that are more efficient than traditional models. The technology is now fairly well understood. But few of these new stoves get beyond the prototype stage because they are ill-matched to people's real needs.

Whether people in a particular place will use a stove does not depend only on the stove's efficiency. A great deal depends on the traditions and economic level of the people expected to use it and on whether they buy their fuel or gather it freely.

Poor people who collect their fuel without payment will rarely be able to invest much in an energy-saving stove. In such places, stoves will have to be extremely cheap if they are to be adopted. Where fuel is bought and sold, people should be more willing to invest in stoves that will save them money.

The higher people's incomes, the more sophisticated the stoves that can be marketed — and the less public-sector involvement or subsidies should be needed. In the 19th century, farmers across the rural United States bought iron woodstoves from manufacturers without the help of UN agencies or foreign universities.

The type of stove, its price and the degree of sophistication required in its manufacture and use must be closely matched to conditions in the area where it is to be used. The process of successfully designing and marketing a stove is similar to that which successful manufacturers of most consumer goods engage in. A stove will not be bought or used unless customers want it and can afford it. A failure to obtain continual feedback from potential users and incorporate the findings into design and extension work has been a prime cause of the poor results of stove programmes.

Designing improved stoves: technical factors

The simplest way to improve the open fire is to put a mud enclosure around it to protect it from draughts. Such a shield can also direct the flow of combustion gases more closely around the pot, increasing the amount of heat it absorbs.

It is difficult to offer simple improvements to the traditional mud or ceramic stoves already in wide use in Asia. But adoption of these traditional models would be an improvement in rural Africa, where open fires are still the norm.

The single-pot stove has a basic limitation: once the gases have escaped past the pot there is no practical way of capturing their remaining energy for cooking. This is why most improved stoves have two or more potholes. Channelling the hot gases past more than one hole can mean more efficient use of their heat. When cooking with just one pot, the other holes can be used to heat water or keep food warm.

Multiple-hole stoves do not work well, however, unless the dimensions and relative positions of the fire chamber, gas passages and other stove parts are correct. A poorly designed stove, or one that has deteriorated severely, may be less efficient than a well-tended open fire.

The draught through a stove must be controlled for the fire to burn properly and the gases drawn past the various potholes. This may be achieved using a hinged or sliding door in front of the firebox, which can be adjusted to control the air flow. But such doors often fall off or are damaged with wear. And those needing light for cooking may leave the door open despite the loss in energy efficiency.

A 'Nouna' stove installed in a courtyard in Naimey, Niger. The child is using a simple shielded fire.

Cooking on a stove with a firedoor also means that wood must be cut into short pieces, adding substantially to the work of the household — and perhaps thereby negating the work savings resulting from reduced fuel needs. In many projects, women have completely stopped using firedoors after a time.

To work well, multiple pothole stoves often need a chimney to help create the necessary draught. This has the added benefit of carrying smoke away from the dwelling. But badly fitted and maintained chimneys, too, can cause a stove to consume more fuel than an open fire. Most chimneys are expensive, often costing as much as the rest of the stove. So a chimney can be a mixed blessing, and installing one so that it performs well is not easy.

Short chimneys discharging indoors at ceiling or eaves level are cheaper and are less likely to cause problems of excessive draught through the fire. These are especially appropriate where smoke is required for curing meat or to protect the roof against termites, as they preserve the benefits of the smoke while avoiding the worst of its discomforts.

The short working lives of mud stoves can be lengthened through the use of a prefabricated ceramic liner for the firebox and other critical parts. The liner can be made to a high standard of accuracy by trained potters, and the task of stove building is reduced to the relatively simple operation of embedding the ceramic liners in mud.

Cement can be added to the building material or applied to the surface of the stove to increase strength and durability. Metal or ceramic reinforcements can be built into the stove to strengthen critical points such as the fire chamber or the edges of potholes. Or the whole stove can be made of brick or concrete blocks, though this tends to increase the price greatly.

The main weakness of metal stoves is loss of heat through poor insulation. Designers have concentrated on methods of insulating the firebox, generally using techniques such as the pottery liner employed in the *Thai bucket*.

A few years back, many stove promoters hoped to disseminate improved models that families could build themselves. More recently, as the technical demands of stove efficiency and durability have become more apparent, emphasis has shifted toward models that can be built by well-trained artisans at low cost.

Savings in fuel

Many stove programmes are operating with little verifiable knowledge of whether they are meeting their fuel-saving objectives. Stove expert Timothy Wood has observed:

> "We still do not know how much wood is saved by using 'improved woodstoves'.... One often hears about the total number of stoves built

by this or that project. It would be more useful to know how much wood is actually being saved as a result of these stoves. Surprisingly, most project managers *do not know*, and many do not seem very interested."

Particularly little is known is about the sustained performance of both stoves and their users. Most consumption surveys are carried out soon after the stoves have been installed, when their energy efficiency is at its maximum. This is also the time when the new owners are most self-conscious about fuel use and aware of outsiders' interest in their cooking methods, and hence most likely to tend fires carefully. Researchers observing families cooking over open fires in Upper Volta found that wood use declined by 25% over the course of two weeks of observation — without any change in technology.

Most of the few surveys that have been carried out report some savings in fuel consumption in the intitial period after stoves have been installed. But some have actually found increases in consumption. Others have been quite inconclusive.

Almost every one of these surveys has been based on samples too small for valid statistical analysis or contained other flaws that undermined their reliability. Some rely on hearsay, others on oral responses from women about past and present wood consumption — neither a scientifically reliable indicator.

Blithe claims that new stoves are reducing wood consumption by half are simply unsupportable. Less dramatic savings could still provide ample justification for stove-promotion activities, particularly when the health benefits are taken into account. Unfortunately, too few facts are available about the actual impacts on fuel consumption of most existing programmes to permit judgements about their value.

Limits to national energy savings

High hopes have been placed on stove programmes as a means of saving energy on a large scale at a national level. Work carried out in Malawi by energy economist David French casts doubt on whether such savings can ever be significant. Similar conclusions are likely to apply elsewhere.

In Malawi, a new mud stove tested in a laboratory saved 50% of the fuel used by an open fire to cook a standardised meal. These laboratory savings fell by half with newly built stoves in the field, bringing the fuel savings down to 25%.

Cooking accounts for about 60% of domestic woodfuel use in Malawi. The rest is for heating, lighting, fish and meat drying, beer brewing, water heating and other cooking tasks that would not be carried out on the new mud stove for one reason or another. Thus the maximum possible saving from the stove programme drops to 15%.

Further adjustments need to be made for the deterioration of mud stoves over time, for periods of hot weather when women prefer to cook outdoors (the mud stoves are not portable), and for times when the heat of an open fire is desired. Taking these into account, the average family reduces its annual wood consumption by about 7.5%.

If half the rural population adopted the new stoves, the maximum savings would amount to 3.8% of the wood used by rural households. Since only 40% of the national wood demand in Malawi is attributable to rural households, the final saving is just 1.5% of the country's total annual wood use.

These figures are not immutable. In particular, better stove designs could produce better long-term fuel efficiency. Also, the potential benefits of improved urban stoves are not included. But the point remains: even under quite optimistic assumptions about stove efficiency and performance, new stoves are unlikely to make a major dent in national wood demands.

The lack of durability of many stove designs imposes further constraints on national-level impacts. Timothy Wood has applied the concept of 'half-life' to mud stoves as a means of estimating the cumulative effect of a stove programme over a number of years. The half-life of a radioactive isotope is the length of time taken for the radiation of any quantity of the isotope to fall to half its original level. The decay of each individual atom is not predictable, but that of the total number is.

Timothy Wood suggests that though the life of any one stove cannot be predicted, the rate at which a large number of stoves deteriorates can. The mud stoves being promoted in West Africa, he estimates, have a half-life of between one and two years. If a project involves a constant stove output of, say, 1,000 per year, and the half-life is one year, the maximum number of stoves ever in use will be 1,999. Even if the half-life is quadrupled to four years, the maximum number of stoves in use — assuming a constant level of production — only increases to 6,276.

Thus the impact of any programme will remain very limited as long as it depends on the efforts of outside promoters. The large-scale diffusion of stoves can take place only through a continually expanding programme. In practice this means that the stove model introduced must be one that local entrepreneurs or users themselves are willing to build without outside assistance. The importance of achieving greater stove durability is also underscored.

The role of stove programmes

Some of the more grandiose goals that underlay early stove programmes have been proven to be overambitious. It is clear that stoves provide no answer to the enormous problems of deforestation and rising energy demands facing many developing countries. To the extent they actually cut family wood use, however, they can save families money or time. And from a national point of view they could reduce plantation requirements to a modest degree.

There is no doubt that when stoves are properly designed and fit the local cultural context they can bring many other benefits as well to individual families. Stoves often make cooking quicker and more convenient. They reduce the mess and dirt caused by an open fire and can eliminate or cut down the nuisance and health hazards of smoke. Children are burned less often. In short, improved stoves mark an advance in living conditions.

As societies advance economically, people always adopt stoves at some stage. Promoting improved stoves is really an effort to hasten something that is part of more general development. But stove activities need not be closely linked to forestry projects or run by forest departments. Programmes designed to aid small businesses, to improve the home environment, or to raise family income also provide logical starting points for stove promotion efforts.

Chapter 10

Strategies for the future

Projections of fuelwood supply and demand yield an unpleasant picture. To be sure, local situations differ dramatically, and everywhere both supply and demand patterns will change in response to events. But we can describe with confidence some of the trends that are likely to unfold in many regions.

The number of people requiring fuelwood for home cooking and heating will rise, especially in rural areas. Despite the increased difficulty and expense of obtaining wood, alternatives — whether fossil fuels or renewable technologies such as biogas plants and solar devices — will remain too costly or impractical for huge numbers. At the same time, the natural sources of fuelwood will be further reduced: cultivation will continue to eat into forest areas, while fuelwood collection, grazing, and other forces will deplete non-forest trees and shrubs.

As wood resources decline, so will the welfare of many people — but especially the poor. Urban fuelwood and charcoal prices will rise; those able to will switch to kerosene or other alternatives, those unable to will scrounge for scraps of burnable material and cut back on cooking.

In the countryside, wood will increasingly be commercialised, providing some of the poor with a chance to earn money but placing unwelcome burdens on many more. Better-off rural residents will grow wood on their farms, buy wood, or switch to kerosene, bottled gas, or other alternatives. Landless and small-scale farm families will walk farther and longer collecting wood until supplies are too far away. Then they will move down the energy ladder, turning to dried dung, crop residues, inferior shrubs, roots and leaves. Wood theft and conflicts over access to fuelwood and crop residues will become more frequent.

The rural response

A combination of possible measures can help ease, if not eliminate, rural fuelwood scarcity. They involve a much wider range of activities than plantations, and generally serve a multitude of ends. Popular participation in the design of tree growing programmes and selection of species is a must, and planners may have to accept uncertainty about the extent to which new planting will directly increase fuelwood supplies. Villagers will nearly always be better judges than outsiders about what the priorities for use of their land ought to be.

Where social conditions allow and land is available, village woodlots and other planting activities on common lands can help increase the supply of forest products. However, in many Third World countries, rural plantations are unlikely to meet more than a fraction of future fuelwood needs. Progress in two other areas — increased tree growing on private farmlands and improved management of existing forests — is also vital.

As remaining common lands lose their stock of trees, those farmers not already doing so may have little choice but to extract more household fuel from their own land. Tremendous untapped potential exists to grow more fuels on farms through such measures as the planting of nitrogen-fixing trees around fields, the incorporation of trees into cropping systems and home gardens, and commercial farm forestry. (Obviously, efforts to encourage more tree growing by farmers must include special steps to assist smaller farmers and to prevent the negative social impacts of some past programmes.)

The potential for burning crop residues for fuel without undermining

Mark Edwards/Earthscan

Carefully-planned rows of trees in the Majia Valley, Niger, provide windbreaks against erosion and can increase millet yields by 25-30%. Villagers will divide the profits from fuelwood sales when the trees are pruned.

agriculture deserves urgent scientific study. In some cases, the extra organic produce associated with agricultural intensification could well provide a satisfactory solution to the fuels problem. The use of residues for cooking will seldom be cost-free; but in some circumstances it might be less costly and much more feasible than devoting scarce lands to tree plantations.

While the social obstacles are formidable, improving the management and sustainable wood output of existing forests is another important and relatively neglected option. Many forests in developing countries are in theory protected from grazing and wood cutting, but are in practice grossly overgrazed and overcut.

Forest departments almost everywhere have a history of confrontation with local populations. If, instead of solely concentrating on forest protection, they can treat forest reserves as dynamic resources meant to serve local communities, both forests and people could be better off. If the cooperation of nearby people can be secured, forests can be managed to increase wood output at a far lower cost than that of plantations, and the many non-wood values of forests will be preserved as well. It would in any case be folly to accept the continuing degradation of forests as an unalterable trend.

The urban challenge

The continued reliance on wood or charcoal by urban populations presents special problems but also special opportunities for solutions. These concentrated markets demand huge amounts of wood, and consumers have little chance to grow their own. On the other hand, city residents generally have higher incomes than rural people and the fuelwood economy is already monetised.

In many regions, rising urban fuelwood prices should present greater opportunities for profitable commercial tree growing. If the economics are right, governments can promote various measures such as intensive government-run plantations, farm forestry, regulated cutting of forest reserves, and even productive management of greenbelts and city trees. Where charcoal is the norm, efforts can be made to establish sustainable sources of wood and to increase the efficiency of charcoal conversion and transportation.

With the rise of commercial fuelwood markets, millions of poorer people living around cities and in distant wood-supplying areas have come to rely on the wood trade for part or all of their livelihoods. Any schemes for bolstering urban wood or charcoal supplies should strive to incorporate such people, giving them a chance to earn more income in a newly sustainable wood business. The thoughtless establishment of new plantations and marketing systems could displace and further impoverish some of the very people supposedly being helped.

In some situations, commercial firewood growing may simply be uneconomical; even sharply higher firewood prices will not produce profits equal to those from growing food, export crops, poles or pulp. Because uncontrolled deforestation exacts a social price that is not included in any individual farmer's financial calculations, a case may well exist for public subsidies to encourage commercial fuel growing. But wood as a commercial product must stand on its own against the economic alternatives. When the point is reached where wood can no longer be extracted as a 'free' good from the environment, the economics of its production and use must be compared with those of alternative cooking fuels.

Already, kerosene and bottled gas are competitive with firewood for family cooking in many cities. In some cases the greater displacement of wood by fossil fuels has been prevented only by government restrictions on imports or on the share of refined oil devoted to the household market. These restrictions are often imposed because of foreign exchange shortages. Governments have, perhaps unknowingly, traded the invisible social costs of deforestation for the more visible costs of fossil fuel imports. (Elsewhere kerosene is subsidised, creating a different set of problems but reducing wood demands.)

If a true accounting of the costs and benefits of wood growing shows it to be uncompetitive with alternatives, countries may have to resign themselves to devoting more fossil fuels and more foreign exchange to meeting household fuel needs. It is of course tragic for a poor country to have to devote a greater share of its scarce funds to consumption rather than productive investment. But this is the tragedy of poverty and underdevelopment. No good can come from exempting commercial wood from normal economic accounting — from concealing the real costs of growing it today because it seemed to be free yesterday.

Fuel for cooking and home heating is a basic human need. Critics of inequitable agricultural strategies have called for governments to put 'food first' — to ensure above all that their populations can feed themselves. Likewise, governments must put 'fuel for cooking first'. Often in the past, fossil fuel imports have been devoted to the demands of the small modern sector while the elementary fuel needs of the urban majority have been met at the expense of biological capital and of rural villagers whose dwindling wood supplies are drawn to city markets.

The wider adoption of a non-renewable fuel whose international trade is subject to politically induced disruptions is not, of course, without risk. But even if all the Third World's urban residents switched to kerosene, their household consumption would not add up to more than a few per cent of global oil use. And, unless the rich countries engage in another massive increase in oil consumption, which appears unlikely, world supplies at near the current annual level should be available for several decades. While we are not urging a thoughtless shift to fossil fuels, they surely deserve serious consideration in many countries as a medium-term answer to what may otherwise be an intractable urban fuelwood problem.

As forest depletion becomes extreme, new patterns of household fuel supply are inevitable. No changes will be cost-free. The challenges are to minimise the costs and to prevent them from being borne disproportionately by the poor or, via further environmental degradation, by the generations to come.

As urban firewood and charcoal prices rise, more efficient stoves should become more attractive. Consumers will want them if they save money within a relatively short time and do not pose any serious disadvantages compared to traditional cooking methods. Governments will want to encourage their adoption if wood savings through improved stoves come more cheaply than the equivalent amount produced in plantations. But improved stoves will eventually have to stand the test of the marketplace. Logically, their use should take off in cities long before it does in rural areas where wood is only now being commercialised.

Like urban consumers, wood-burning industries have often enjoyed a free fuel ride at the long-term expense of forests and villagers. Governments can force them to develop their own sustainable sources of wood or, if this is not feasible, to switch to alternatives. If an industry cannot pay the true costs of its production and remain competitive, then its gradual demise may be in order. Or if societies choose to subsidise an industry by providing wood at below its real cost, this should be done with eyes open.

Focus on the poor

More complete commercialisation of woodfuel will in itself help ease wood scarcity: new wood growing will be elicited, wood conservation will be encouraged, and some consumers will switch to newly competitive alternatives. But there is a major problem with all the measures suggested so far. The energy plight of the Third World's underclass, rural and urban, will continue to worsen.

The growing legions of landless labourers and farmers with small, marginal plots will not be able to grow their own fuel. To the extent that the needs of better-off farmers and urban consumers are met through new tree planting or fuel switching, pressures on common lands may be reduced and the amount of fuel they provide the poor increased. But the unfolding commercialisation of wood and even crop residues will put unbearable pressure on the meagre, often declining real incomes of the rural poor.

Special programmes to help the rural poor meet their basic household fuel needs must be the heart of any country's community forestry efforts. Where village woodlots or roadside plantations are attempted, organisers must ensure that the poor are involved in planning, given priority for jobs created, and given a good share of the proceeds. Those who lack the land to grow their own forest products can be given preferential access to forest reserves or receive allocations of public land for tree growing.

But all such measures together cannot fully ease the burdens that fuel scarcity and rising wood prices will impose on the poor. More scavenging for flammable materials and less cooking, possibly with adverse nutritional consequences, are in prospect for those on the bottom.

The poorest urban people, too, will be left in the cold by rising wood prices. Those who can will squeeze out a higher share of their incomes for wood or kerosene and perhaps will eventually be helped to adopt fuel-saving stoves. But many may do less cooking themselves, instead buying food from vendors whose constant commercial cooking is more fuel-efficient.

Beyond fuelwood

Some analysts have calculated how much tree planting would be necessary in the coming years to bring the supply of fuelwood into balance with the projected demand. The results have been shocking, indicating a need for anywhere from a fivefold to, in some parts of Africa, a twentyfold or greater increase in the area planted. These increases are for many countries far beyond any realistic possibility, even assuming much greater local commitment and international support than in the past.

Although forest departments in many countries are now gearing up for massive increases in private and communal tree planting, the many social, bureaucratic, and resource constraints on community forestry reviewed earlier in this book will not disappear — and many of the trees planted in successful programmes will be grown for purposes other than fuel production. Even if they are extraordinarily successful, new plantations will supply nowhere near the amounts of fuelwood called for by demand projections. Nor can improved cookstoves, even if they are adopted far more widely than they have been to date, significantly alter the fuelwood outlook.

This is not to say that community forestry and cookstove programmes are not urgently needed. Rather, we suggest that in many countries it may be impossible to 'solve' the fuelwood problem through tree growing alone. Fuelwood scarcity must be seen and attacked as one element of the broader web of agricultural, forestry, energy and economic challenges.

Activities that contribute to the solution of any of these challenges can, directly or indirectly, help alleviate fuelwood scarcity. Growing more fodder, for example, can mean more dung production, better fertilised fields, and the increased production of organic matter including fuels. Growing poles for the market can, by producing new income, improve a family's ability to buy a more efficient stove or switch to new fuels. But it is an illusion to think that the fuelwood problem can be attacked and conquered in isolation from the more general problems of land use and underdevelopment.

The firewood crisis is but one aspect of the crisis of underdevelopment and powerlessness afflicting the poor of the Third World. Several hundred million people whose incomes are stagnant or falling are seeing their traditional sources of cooking fuel depleted and commercial wood or other fuels priced out of reach. Many more people with higher or rising incomes also face fuelwood-related hardships, and their continued wood use will contribute to the overcutting of forests. But their household fuel problem is much more tractable than that of the underclass.

For those who are really poor, the depletion of formerly free firewood supplies means that fuel joins food, water, and housing on the list of basic needs that are satisfied inadequately and with great trouble. These people are certainly aware of new burdens, but, given their other pressing needs and the constraints on tree growing, cannot be expected to place especially high priority on solving the fuelwood problem.

Social reforms and development that gives those on the bottom a chance to improve their lives would help relieve firewood scarcity in several ways. As family welfare improved and more children survived, family planning would become more thinkable and future demands for wood would be reduced. Equitable economic progress could also make possible the stabilisation and productive management of forest areas; when the rural landless have better opportunities to make a living, they will not cut down forests in desperate efforts to feed their families. And as their incomes rise, people become able to buy wood from sustainable sources or to use alternative fuels.

As with hunger and disease, the fuelwood crisis will not be eliminated through technical measures alone, necessary as these are. People have difficulty cooking their food and heating their homes because they are poor. There is no escaping the social challenges of underdevelopment.

EARTHSCAN PAPERBACKS

A Village in a Million by Sumi Chauhan   1979   £2.00/$5.00

Climate and Mankind by John Gribbin   1979   £2.00/$5.00

Antarctica and its Resources by Barbara Mitchell and Jon Tinker   1980   £2.50/$6.25

Mud, mud — The potential of earth-based materials for Third World housing by Anil Agarwal   1981   £2.50/$6.25   Also in French & Spanish

New and Renewable Energies 1 (solar, biomass) edited by Jon Tinker   1981   £2.50/$6.25   Also in French & Spanish

New and Renewable Energies 2 (others) edited by Jon Tinker   1981   £2.50/$6.25   Also in French & Spanish

Water, Sanitation, Health — for All? Prospects for the International Drinking Water Supply and Sanitation Decade, 1981-90 by Anil Agarwal, James Kimondo, Gloria Moreno and Jon Tinker   1981   £3.00/$7.00

Carbon Dioxide, Climate and Man by John Gribbin   1981   £2.50/$6.25

Fuel Alcohol: Energy and Environment in a Hungry World by Bill Kovarik   1982   £3.00/$7.00

Stockholm Plus Ten: Promises, Promises? The decade since the 1972 UN Environment Conference by Robin Clarke and Lloyd Timberlake   1982   £3.00/$7.00

Tropical Moist Forests: The Resource, The People, The Threat by Catherine Caufield   1982   £3.00/$7.00   Also in French & Spanish

What's Wildlife Worth? by Robert and Christine Prescott-Allen   1982   £3.00/$7.00   Also in Spanish

Desertification — how people make deserts, how people can stop and why they don't by Alan Grainger   1982   £3.00/$7.00   Also in French

Gasifiers: fuel for siege economies by Gerald Foley, Geoffrey Barnard and Lloyd Timberlake   1983   £3.00/$7.00

Genes from the wild — using wild genetic resources for food and raw materials by Robert and Christine Prescott-Allen   1983   £3.00/$7.00

A million villages, a million Decades? The World Water and Sanitation Decade from two South Indian villages — Guruvarajapalayam and Vellakal by Sumi Krishna Chauhan and K. Gopalakrishnan   1983   £3.00/$7.00

Who puts the water in the taps? Community participation in Third World drinking water, sanitation and health by Sumi Krishna Chauhan with Zhang Bihua, K. Gopalakrishnan, Lala Rukh Hussain, Ajoa Yeboah-Afari and Francisco Leal   1984   £3.00/$7.00

Stoves and trees by Gerald Foley, Patricia Moss and Lloyd Timberlake   1984   £3.50/$7.00

Natural disasters — Acts of God or acts of Man? by Anders Wijkman and Lloyd Timberlake   1984   £3.50/$7.00

Stoves and trees
by Gerald Foley, Patricia Moss and Lloyd Timberlake

How much wood would a woodstove save, if a woodstove could save wood?

Do improved stoves actually save fuel — in the village as well as in the laboratory — can they help slow down deforestation? Based on Earthscan Technical Report No.2, it looks at the potential for improved stoves and explains why they are so difficult to introduce in practice.   1984   £3.50/$7.00

All Earthscan publications are available from:

Earthscan
3 Endsleigh Street
London WC1H 0DD, UK.

Earthscan Washington Bureau
1717 Massachusetts Avenue NW
Washington DC 20036, USA

Earthscan Technical Reports

No. 1: Biomass Gasification in Developing Countries
by Gerald Foley and Geoffrey Barnard

A detailed assessment of the current status of gasification technology and its prospects in developing countries. Based on a study for the World Bank, the report provides a comprehensive analysis of the economics of gasifiers, a discussion of potential applications and a review of research and commercial developments worldwide. The report also contains background details of the history, chemistry and technology of gasification.   1983   £10.00/$20.00

No. 2: Improved Cooking Stoves in Developing Countries
by Gerald Foley and Patricia Moss

Analyses the role of improved cooking stoves in the wood-scarce areas of the developing world, and reassesses the rationale for programmes in the light of experience to date. Topics covered include the traditional use of stoves and open fires, patterns of fuelwood consumption and their relation to deforestation, and design and promotion of improved stoves, and the potential and limitations of stove programmes.
1983   £10.00/$20.00

Farm and Community Forestry
by Gerald Foley and Geoffrey Barnard

Over the past decade, farm and community forestry has emerged as one of the principal responses to the problems caused by the widespread loss of trees and forest cover in the developing world. Programmes to promote tree growing by rural people have been launched in more than 50 countries.

These have shown that under the appropriate conditions farm and community forestry can be very effective. It can provide substantial benefits both at an individual and a community level.

But it is also clear that it is an approach which is far from easy. Some programmes have failed completely, while others have had unexpected, and sometimes undesired, side-effects.

This report provides a systematic appraisal of the experience to date with community forestry. It describes the main lessons that have been learnt, and analyses the factors which determine the scope and impact of programmes under local conditions.   1984   £10.00/$20.00